AUDIOVISUAL TELECOMMUNICATIONS

BT Telecommunications Series

The BT Telecommunications Series covers the broad spectrum of telecommunications technology. Volumes are the result of research and development carried out, or funded by, BT, and represent the latest advances in the field.

The series includes volumes on underlying technologies as well as telecommunications. These books will be essential reading for those in research and development in telecommunications, in electronics and in computer science.

1. *Neural Networks for Vision, Speech and Natural Language*
   Edited by R Linggard, D J Myers and C Nightingale.
2. *Audiovisual Telecommunications*
   Edited by N Kenyon and C Nightingale
3. *Telecommunications Local Networks*
   Edited by W R Ritchie and J Stern
4. *Digital Signal Processing in Telecommunications*
   Edited by F Westall and A Ip

AUDIOVISUAL TELECOMMUNICATIONS

Edited by

N D Kenyon

and

C Nightingale

BT Laboratories, Ipswich, UK

CHAPMAN & HALL

London · Glasgow · New York · Tokyo · Melbourne · Madras

Published by Chapman & Hall, 2–6 Boundary Row, London SE1 8HN

Chapman & Hall, 2–6 Boundary Row, London SE1 8HN, UK

Blackie Academic & Professional, Wester Cleddens Road, Bishopbriggs, Glasgow G64 2NZ, UK

Van Nostrand Reinhold Inc., 115 5th Avenue, New York NY10003, USA

Chapman & Hall Japan, Thomson Publishing Japan, Hirakawacho Nemoto Building, 6F, 1–7–11 Hirakawa-cho, Chiyoda-ku, Tokyo 102, Japan

Chapman & Hall Australia, Thomas Nelson Australia, 102 Dodds Street, South Melbourne, Victoria 3205, Australia

Chapman & Hall India, R. Seshadri, 32 Second Main Road, CIT East, Madras 600 035, India

First edition 1992

© 1992 British Telecommunications plc

Printed in Great Britain by St Edmundsbury Press Ltd, Bury St Edmunds, Suffolk

ISBN 0 412 45800 4   0 442 30879 5 (USA)

Apart from any fair dealing for the purposes of research or private study, or criticism or review, as permitted under the UK Copyright Designs and Patents Act, 1988, this publication may not be reproduced, stored, or transmitted, in any form or by any means, without the prior permission in writing of the publishers, or in the case of reprographic reproduction only in accordance with the terms of the licences issued by the Copyright Licensing Agency in the UK, or in accordance with the terms of licences issued by the appropriate Reproduction Rights Organization outside the UK. Enquiries concerning reproduction outside the terms stated here should be sent to the publishers at the London address printed on this page.

The publisher makes no representation, express or implied, with regard to the accuracy of the information contained in this book and cannot accept any legal responsibility or liability for any errors or omissions that may be made.

A catalogue record for this book is available from the British Library

Library of Congress Cataloging-in-Publication data available

Printed on permanent acid-free text paper, manufactured in accordance with the proposed ANSI/NISO Z 39.48-199X and ANSI Z 39.48-1984

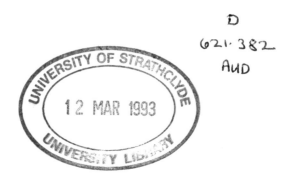

Contents

	Contributors	vii
	Preface	ix
1	**Audiovisual telecommunications** N D Kenyon	1
2	**Broadcast TV systems** S Searby and P M Elmer	27
3	**Television of improved quality** B R Patel	45
4	**A NICAM digital stereophonic encoder** G T Hathaway	71
5	**Videophony** I Corbett	85
6	**Multipoint audiovisual telecommunications** W J Clark, B Lee, D E Lewis and T I Mason	100
7	**Infrastructure standards for digital audiovisual systems** N D Kenyon	113
8	**Practical low bit-rate codecs** M W Whybray and M D Carr	133
9	**Variable bit-rate video coding for asynchronous transfer mode networks** D G Morrison	156
10	**Standardization by Iso/MPEG of digital video coding for storage applications** D G Morrison	177
11	**Model-based image coding** W J Welsh, S. Searby and J B Waite	194
12	**Head boundary location using snakes** J B Waite and W J Welsh	245

13	**Image data compression using fractal techniques** J M Beaumont	266
14	**Artificial neural nets for image processing in visual telecoms: a review and case study** R A Hutchinson	299
	Index	337

Contributors

J M Beaumont	Image Processing, BT Laboratories
M D Carr	Video Systems, BT Laboratories
W J Clark	Video Systems, BT Laboratories
J M Corbett	Image Processing, BT Laboratories
P M Elmer	TV Networks and Video Applications, BT Laboratories
G T Hathaway	TV Networks and Video Applications, BT Laboratories
R A Hutchinson	Image Processing, BT Laboratories
N D Kenyon	Principal Engineering Advisor, Broadband and Visual Networks, BT Laboratories
B Lee	Video Systems, BT Laboratories
D E Lewis	Video Systems, BT Laboratories
T I Mason	Image Processing, BT Laboratories
D G Morrison	Video Systems, BT Laboratories
D J Myers	Image Processing, BT Laboratories
C Nightingale	Image Processing and Video Coding Standards, BT Laboratories
B R Patel	TV Networks and Video Applications, BT Laboratories
S Searby	TV Networks and Video Applications, BT Laboratories
J M Vincent	Image Processing, BT Laboratories
J B Waite	Image Processing, BT Laboratories
W J Welsh	Image Processing, BT Laboratories
M W Whybray	Image Processing, BT Laboratories

Preface

The concept of combining images and sound in a single communication signal began with the introduction of broadcast television over 50 years ago. However until recently the cost of transmitting a TV signal over any but the shortest distance has been so high that most other uses of moving pictures were out of the question. Only still images could be transmitted on the general telephone network, using facsimile and slowscan TV techniques, but the process was irritatingly slow. Cameras and storage media were also too expensive for many purposes.

Studies carried out by BT and other telecomms operators in the 1970s, looking into audiovisual applications, concluded that not only had technology still a long way to go in reducing costs, but there were also human factors and social obstacles to be overcome. Furthermore, applications such as videophony and audiovisual databases would require the widespread availability of terminals: such a condition would be hard to reach in the absence of strong market pull, and indeed there was scarcely even an awareness of the technical possibilities. For the time being attention became focused on closed-user-group applications – surveillance, picture libraries, teleseminars and videoconferencing. Workers at BT Laboratories were proud to be among the first to produce equipment which would transmit digital surveillance and database pictures over the telephone lines and, in 1981, the first 2 Mbit/s monochrome videoconference pictures.

Since that time there have been great advances in technology – the great surge in consumer electronics has brought the cost of colour cameras and moving-picture storage media to quite modest levels; speech and image processing algorithms have improved to the point where audiovisual signals can be compressed to very low digital transmission rates; integrated circuits are now becoming available to achieve such compression economically; and finally the emergence of digital local area networks and ISDN obviates the need for a transmission medium particular to vision signals.

Throughout this period BT Laboratories played a leading role, in basic research, in product development (particularly for videoconferencing) and in the important matter of standardization.

This book presents much of this work by bringing together chapters on a number of important facets of audiovisual communications. Attention is given in the first half of the book to applications and system aspects, as befits a telecommunications company. At the time of writing, changes are taking place in both entertainment and business uses of television, changes which will undoubtedly continue into the next century. Technological advances – digital and high-definition TV, so-called 'broadband' networks, the rise of the liquid crystal display – can be mapped with far greater confidence than their commercial repercussions.

In the latter half of the book we move on to discuss image processing in this telecommunications context, from the currently practicable real-time machines entering the marketplace to the research frontier which will open up new applications possibilities in the future.

The principal target of audiovisual development work in the telecommunications laboratories has been an infrastructure of technology and standards as a platform on which a wide range of plausible marketplace applications can be built; there are now encouraging signs of demand for such applications from the general business public, and we may be confident that there will be strong growth in the next ten years, as the public learns to see, as well as speak, at great distances.

N D KENYON
Principal Engineering Advisor, Broadband and Visual Networks
BT Laboratories

1

AUDIOVISUAL TELECOMMUNICATIONS

N D Kenyon

1.1 INTRODUCTION

A wide variety of services can be envisaged in which speech or text communication between two or more terminals is augmented by other, visual forms of communication, such as documents, diagrams or other forms of still image, moving images, control signals and so on. On the classical analogue telephone network such services are difficult to achieve in a convenient and cost-effective way.

The introduction of a visual element into telecommunications is made possible by the great advances in technology made since the 1970s, advances in integrated-circuit memories and processes as well as in television cameras and displays. In many applications, however, this visual element is not the principal mode of communication. The term 'audiovisual' is chosen to convey the concept of a system which better matches the communication needs of users by transmitting a combined set of information formats, destined for the eyes as well as ears of the recipients.

In one sense, the earliest telecommunications were visual (smoke signals, semaphore) because light stimuli can travel a considerable distance — not to mention their speed. Even the first electric telecommunications were visual (telegraph) but this had more to do with the sensitivity of the eye to small stimuli, and the fact that incoming signals could not be recorded in audible form.

However, in both these cases the information transmitted did not have the significance of an image, but of other information in coded form (e.g. the binary representation of characters). However, in present-day terms, visual telecommunication is taken to involve the transmission of image information, the great practical difference being that images potentially contain vastly more information than any other form of communication.

The word 'potentially' is used advisedly — if every element of an image is considered to contain information, then a modest field of view contains an enormous magnitude of information. If however 'information' refers to those aspects of an image which convey meaning to the viewer, then the content may be rather low (e.g. 'a tree' — if the viewer is not concerned with its species or its exact shape).

Assuming that the human eye can resolve down to 2 minutes of arc, it can take in something like a million picture elements (pixels) without moving; moving the eyes (but not the head), the field of view is almost an order of magnitude greater. To represent the intensity and colour of any one pixel is known to take about 12 bits, in binary terms. Thus, to code in a simple way a static scene, something like 100 million bits of information are normally required.

Total changes of this scene (apart from closing the eyes or switching the lights off) occur due to head movement, taking place in a time ranging typically from 0.2 to 5 s. Within a scene itself however, change may be much more rapid — the most obvious case is that of watching a film in which there is a cut from one scene to another. It is clear that the traditional 24, 50 or 60 images per second of film and television media are required to cope with sudden changes and rapid movement; but in no way should the information flow be interpreted as these multiples per second of the number of pixels.

Integral images can be transmitted unchanged over several tens of metres, by optical fibre or as a beam in free space, all the information travelling in parallel. However, for telecommunications purposes only one time-dependent variable is allowed — samples are therefore taken sequentially across the whole image and transmitted to line, repeating this process as often as possible. Such a simple scheme is able to cope with unpredictable sudden changes in the image, and with almost random spatial detail over a quite limited field of view. For a conventional television image the signal flow, in binary terms, is of the order of 100 Mbit/s. This poses no great problems for a single moving picture over limited distances, but for longer distances the cost is not trivial.

For transmission-cost reasons alone it is necessary, for some purposes, to reduce the binary flow, by three or even four orders of magnitude; this can be done either by removing redundancy from the images or by simplifying them in some way, or both. In practice, redundancy-reduction techniques generally involve some compromise to the image, be it simplification or other distortion, and it is true to say that the higher the binary compression the more compromised the resultant image will be. Thus transmission-cost reduction is achieved, but if the image is no longer fit for the purpose this becomes pointless.

It can be concluded that visual telecommunication is all about compromise between what is really seen and what actually needs to be seen for the intended application, such compromise being necessary because of costs, technical limitations, or both. Before going on to consider the applications, we look briefly at the current state of image processing for data compression and the compromises involved.

1.2 IMAGE PROCESSING

1.2.1 Content of a TV image

Figure 1.1 shows a square photograph of a well-known personality at a number of resolutions ranging from 8×8 pixels to 512×512; viewed from a sufficient distance, the image is recognizable even at 16×16 because the subject is so familiar; the quality improves greatly as the resolution increases. The rectangular image of conventional TV contains, in sample terms, the equivalent of 720×576 pixels [1], though subjectively the resolution is somewhat lower because of the characteristics of the cathode-ray tube; high-definition television (HDTV) is about twice this resolution in both directions. Table 1.1 gives a rough quality statement for various resolutions.

Table 1.1 Quality of an image for various resolutions

	Quality v Resolution
>1440 (h) $\times$ 1152 (v)	35 mm slides, high definition TV
720×576	TV studio standard
360×288	videocassette or surveillance quality
180×144	small-picture videophone
90×72	poor quality, even for small pictures
45×36	barely recognizable

4 AUDIOVISUAL TELECOMMUNICATIONS

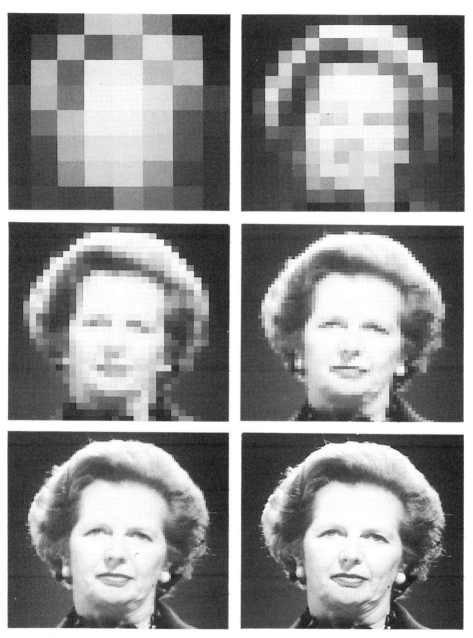

Fig. 1.1 Recognizability and quality of an image depend on both the resolution and viewing distance.

For studio-quality television the pixels are quantized linearly to 256 levels, requiring an 8-bit code; Fig. 1.2 shows that a difference of 2 levels in 256 is quite perceptible, though the case shown is extreme. If the same scale is used for the colour-difference components, it is found that the spatial resolution does not need to be so great — half the luminance value is sufficient; for example, a videoconference picture of 360×288 luminance pixels requires 180×144 samples of each of the two colour signals.

Fig. 1.2 Perceptibility of a small difference in luminance: the central rectangle is 2 parts in 256 lighter than the surrounding area.

Using the above figures the 'fundamental' content of a conventional TV frame to European standards is about 5 Mbits, and at 25 frames per second the transmitted bit rate would be 125 Mbit/s. On the NTSC standard used in other parts of the world the spatial resolution is lower, but as the frame rate is 30 per second the net bit rate is almost identical.

Of course the human eye/brain system cannot possibly cope with a real information flow of this magnitude; consider for example the effect of viewing 25 totally different frames in a second, or a screen of 720×576 pixels being black/white at random. In practice there is strong correlation both temporally and spatially, and thus a great deal of redundancy. Where there is an economic incentive to reduce the bit count, for transmission or storage, removal of some of the redundancy is reasonable.

1.2.2 Quality

The eye/brain system is quite sensitive in some ways, and yet tolerant in others. It is sensitive to unexpected movement, and so perceives noise from analogue transmission or digital coding artefacts. It is sensitive to distortions in familiar images: since we are familiar with the clarity of directly viewed objects, the soft image formed from a lower-resolution sampling process may be disturbing. Again, we are particularly aware of those parts of images conveying critical information, such as the eyes of another person, or alphanumeric characters. Moreover the eye effectively integrates over a large area, corresponding to a signal time which is long compared with the detail in it.

Tolerance increases where the information used in the brain is not that conveyed by the image: we adjust very well to variations in relative and absolute brightness of image, and would rarely notice if rates of movement were not accurately represented. Often the clarity of background is of little significance.

There is thus considerable scope for reduction of objective quality with little, or at least a tolerable, effect on the perceived quality.

1.2.3 Image compression

For a treatise on this complex subject, see [2]. Here only a brief survey is given of the basic elements of image compression as applied in the commercial video codecs now available, and an introduction to the newer techniques which are the subject of later chapters.

Table 1.2 Methods of bit-rate reduction

1.	Redundancy reduction — areas unchanged from frame to frame — areas which have moved but otherwise not changed — adjacent pixels or blocks of pixels which are the same — shorter codes for the commoner features
2.	Quality reduction — spatial resolution — temporal resolution — size reduction — luminance approximation — feature approximation

Table 1.2 lists the various means by which the content of a coded image can be reduced, classified principally into redundancy reduction and quality reduction. It should be noted that these are not independent: reductions in picture-size or spatial/temporal resolution may be deliberate, whereas efforts to remove redundancy generally result also in approximations and reduction of resolution.

Redundancy reduction techniques assume that there is correlation between neighbouring (in space and/or time) pixels or within blocks. The details of correlation are encoded and also the 'errors' — the differences between the assumptions and reality. To keep the volume of data low, approximations must be used for both, and although chosen to deceive the eye as much as possible these approximations do involve a perceptible loss of quality.

Each frame can be treated either pixel by pixel or by blocks of pixels, where a block might be 8×1, 8×8, 16×16 ... or ultimately the whole picture.

Predictive coding involves the assumption that a value is related to some neighbouring values and may therefore be calculated at the receiver, instead of being transmitted; it is, however, necessary to transmit the 'prediction error' arising from such an assumption, though a relatively crude approximation to the error may suffice.

Consider the case of a single frame. Some of the pixels are coded exactly; these are then used as predictors for others, the latter only being represented, in storage or transmission, by the prediction over error, which is the difference between their real value and the prediction. The derived pixels are in turn used to predict other pixels, and so on. The proportion of exact pixels to those derived maybe as much as half of all pixels, or as little as one per scan line. The process is illustrated in Fig. 1.3, for the simple case in which the first pixel of a string is transmitted exactly, while each subsequent pixel is sent as a difference from its predecessor. Of course there is no gain if 8 bits are used for the prediction error: in practice 3 or 4 bits are sufficient if the difference is quantized non-linearly, the eye being tolerant to the errors thereby introduced. In the illustration, a luminance step from 0 to 141 is transmitted as {sample #1: absolute value 0; samples #2-7: difference values 0, 0, 0, +125, +28, −16, +4}.

In more complex schemes, the prediction may be an arithmetic combination of several pixels, in the neighbourhood of the pixel predicted and preceding it in the processing.

Transform coding seeks to exploit correlation of the pixel magnitudes within a block, by finding another set of magnitudes, many of which will be rather small; the transformed set itself contains no less data in principle, but in practice for most pictures the great majority of the values are infinitesimal or can be discarded without much loss of quality to the eye.

8 AUDIOVISUAL TELECOMMUNICATIONS

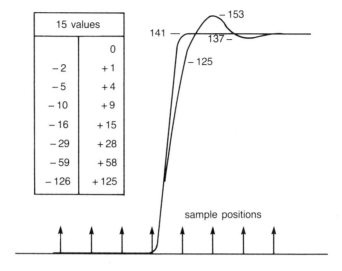

Fig. 1.3 An example of 4-bit difference coding.

Consider first a $[1 \times n]$ block of pixels along a scan line. It is well known that the waveform representing the luminance of any such string of pixels can be represented by the superposition of a number of regular fixed shapes, such as shown in Fig. 1.4(a), with magnitudes suitably adjusted. More commonly the transforms are applied to 8×8 or 16×16 blocks, so the components must have spatial variations in both directions (see Fig. 1.4(b)). In other fields Fourier analysis is often used for the transformation, but in the case of picture coding there are other sets of patterns which give better results: recently the discrete cosine transform (DCT) has become very popular, and in fact this is not so different from the Fourier transform. Essentially, the block of pixels is being represented by a number of spatial harmonics of various amplitudes. The property of these transforms is such that, when the block has been transformed, only a very few of the components have a significant magnitude; the coefficients of these are encoded, while the rest can be discarded.

Until recently, sheer complexity ruled out the DCT; however as VLSI chips became practicable (100 000 gates) and the DCT was incorporated in standardized video coding algorithms (see Chapters 8—10), its popularity has greatly increased, and is now such that VLSI chips are on the market, used in almost all high-compression codecs; great ingenuity goes into efficient implementation of the process by which the few significant coefficients are selected and labelled.

IMAGE PROCESSING 9

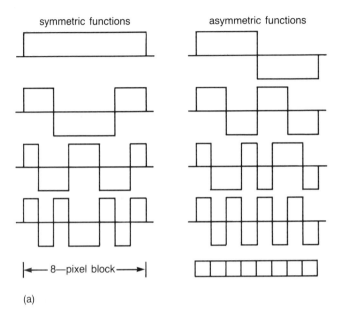

Fig 1.4(a)   The basis of a transformation.

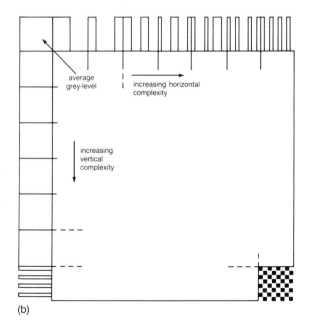

Fig. 1.4(b)   Two-dimensional transform function.

It should be noted that the DCT chips still have their limitation: processing speed must increase with resolution if the full frame rate is to be maintained; thus coding schemes for entertainment-quality television cannot yet be very complex.

Because still pictures do not inherently have a time dependence, the temporal dimension may then be used to achieve a better match to a user's needs: instead of building up the whole picture at a single pass, it may be represented first in approximate form using only the most significant coefficients, filling in the detail subsequently in further stages, a process known as 'progressive updating'. It enables the user to see quickly whether the incoming image is likely to be of interest, and move on to another if it is not.

Interframe coding — both predictive and transform coding can be applied to moving pictures as well as to single images. Anyone who has looked at a strip of movie film will appreciate the great similarity between one frame and the next: clearly this offers a very good opportunity for redundancy removal. At the encoder, frame-to-frame differences are obtained and used to identify the changed areas for further processing by intra-frame coding and/or other methods. Alternatively, a process such as DCT can be applied directly to the difference picture, since coefficient values will all be zero in unchanged areas of the frame.

Movement compensation — for very high compression it is worth taking the process a stage further, by the inclusion of so-called movement compensation: this involves the identification of areas in sucessive frames which appear to correspond (they will rarely be identical) but which have moved; the displacement vector for each such area is calculated. The first-frame data and the displacement vectors are used to derive predictors for the second frame, and then the prediction errors are calculated, these being rather smaller than if the displacement information were not used (Fig. 1.5). Thus each frame can now be represented by the aggregate of the vectors and prediction errors, with addressing to indicate to which areas they apply. The prediction error will be greater where the object has been rotated as well as displaced, and indeed rotation into the third dimension causes new surfaces to appear in the image. Movement compensation is very complex, and only gives significant gain when compression is high; nevertheless VLSI chips are now available and the method is commonly employed in high-compression codecs.

It will be apparent from the foregoing that, where frame differences and movement compensation techniques are employed, the stream of data generated is not constant in time, but rather varies with the amount of motion.

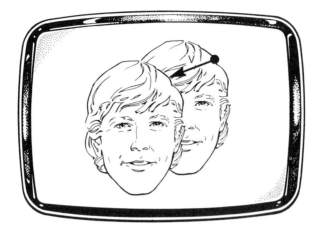

Fig. 1.5 Movement compensation: the displacement vector.

The variation can be smoothed out to a certain extent by buffering, but for a fixed bit rate transmission channel it is usually also necessary to introduce further quality compromise as the amount of motion increases: this is most frequently done by frame-dropping, the perceived effect being that motion becomes jerky. Here again great ingenuity is used in pre- and post-processing to present the best possible image to the user under the constricted circumstances. Chapter 9 considers the case where the transmission bit rate is itself variable.

Movement representation need not be exact provided that it is realistic: for example if a hand movement taking 0.4 s were conveyed at the receiver as taking 0.5 s, the user would probably not notice; however if the hand appeared to flex oddly during its displacement, or if the movement seemed jerky, this would be taken as poor quality. Unfortunately the human eye is rather sensitive to the correspondence between lip movement and speech, a factor which places some strain on current codec performance at low bit rates: although the sound can if desired be delayed by a time roughly corresponding to the video processing delay, the mouth jerkiness caused by frame-dropping is very noticeable. Since people naturally have more body motion when they speak than when just listening, frame-dropping is then more likely and the distortion more marked.

1.2.4 Objects in images

The processing in the foregoing paragraphs has been on a pixel-by-pixel or block-by-block basis, not related in any way to the objects within the image. Blocks of a moving hand, for example, will undoubtedly be found to have similar motion vectors, but there is no recognition of the hand itself — there is no understanding of the change of shape as it moves, nor of the surfaces appearing and disappearing. Human beings recognize familiar object classes very quickly, albeit after years of training. They also recognize human faces and forms very quickly when well-lit and in a familiar environment. Recognition of individual objects may be much slower (for example a particular foreign coin, rather than just any coin).

To match human performance electronically is impractical at present, though remarkable advances are being made (see Chapters 11, 12, 14 following). By the end of the 1990s there will be quite powerful processors which can be trained to recognize a wide range of objects, though perhaps not yet being capable of recognizing individual human beings.

In the field of telecommunications, recognition of object classes could be applied to the improvement of picture quality in high-compression codecs and also to telesurveillance as described in Chapter 14.

1.3. SERVICES AND APPLICATIONS

1.3.1 Classification

A convenient classification of services and applications is shown in Table 1.3; this has been based on standardization work carried out in CCITT [3]. The convention has been used that a 'service' is that which is provided by the technical system, whereas the application is the task for which the system is used. The 'conversational category' is for communications between people, and is therefore generally (but not always) bidirectional, while the other applications are between people and machines, being generally unidirectional or half duplex as far as image transmission is concerned.

Although some degenerate cases have been included for completeness, most of the applications listed are 'multimedia', in that two or more service components are required. In audiographic teleconferencing, for example, wideband (7 kHz) speech is to be preferred, and the means to transmit facsimile, still-picture television (SPTV) or computer-screen material when appropriate; a good system will include conference management aids such as the means to request and to allocate permission to speak.

SERVICES AND APPLICATIONS

Table 1.3 Classification of services and applications

Category	Configuration	Service	Applications examples	Image importance
CONVERSATIONAL Real-time communication between people	Bidirectional, symmetric	Telephone Tp conference Wideband speech Audioconference Audio + still-image ditto, conference Videoconference Videophone ditto, conference	General General General General Medical, training, fault diagn. News, design ... General General General	0 0 0 0 8—10 2—10 4—9 2—6 4—9
	unsymmetric	Teleseminar	Multi-location conferences, auctions, surgical operations, university campus system	10
DISTRIBUTION From one or few machines to many people — perhaps also to machines for local storage	Unidirectional	Radio broadcast TV broadcast TV narrowcast Datacast	Entertainment ditto Intra-business communication Lectures; horse-racing share information Database update	0 10 10 0—1 0—10
	Bidirectional, symmetric (control return)	Switched-star	Entertainment with choice	10
DATABASES (DB) From a few people to machines and then from these to many more people AND MESSAGING (M) Non-real-time communication between people, using machines for storage, routeing, etc.	Bidirectional asymmetric (opposite directions) for storage and retrieval)	Alphanumeric + graphic DB/M Speech input/output DB/M Picture DB/M, keypad control Audio and/or video library (DB) Intelligent DB	General, timetables News, sports, results, pop records, games Correspondence Mail-order catalogues Spare parts identification Training/education Entertainment TV with choice Advertising industry Identify persons: terrorists etc Identify stolen/found property Search for similar symptoms Individualized training	0 0 0—10 5—10 5—10 3—9 10 8—10 8—10 5—10 0—10 3—9
SURVEILLANCE From machine to person (or to other machine)	Bidirectional, asymmetric (control return)	Passive surveillance Active surveillance	Safety/security: areas where there is normally movement Security: detection of movement where there is normally none Recogn. of persons, vehicles ...	10 5—10 10
PROCESSING & TRANSACTION Between person and machine (sometimes involving 2 or more machines)	Bidirectional, symmetric	Transactions Format conversion Remote control Image processing	Voting, EFTPOS Home banking Electronic ordering Speech to text, and vice versa Language translation Telecommand of home equipt. Editing, special video effects Correlation, feature extraction	0 0 0 0 0 0 0—10 0—10
INFO. TRANSFER Between machines	Bidirectional, asymmetric	Facsimile, videomail Data	Correspondence, etc. Remote publishing Data back-up	0-10 0 0

As another example, consider remote training applications: here again high-quality speech is preferred. The high transmission cost of moving-picture television (MPTV) may not be justified; still images may be used as for audioconferencing. Preferably, two picture stores are used at the receiver, so that while the lecturer is discussing one slide the next may be in course of transmission. An X-Y device is indispensable, for pointing at the visual display — it would clearly be undesirable to retransmit the whole image just to include a pointer; some lecturers prefer to sketch or write as they speak, in which case a telewriter or electronic blackboard is necessary. The visual material need not be in hard-copy form even at the lecturer's terminal: he may call on a database for stored images or text/graphics and perhaps also recorded speech.

Figure 1.6(a) shows a lecture terminal incorporating several facilities, including 'soft facsimile' where a document is picked up by a fax scanner and displayed on a high-resolution monitor.

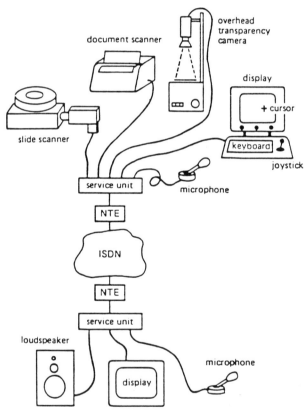

Fig. 1.6(a)   Lecturer's terminal (above) and remote unit.

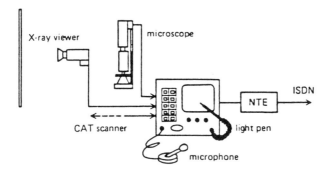

Fig. 1.6(b) Medical terminal.

Conferencing and training aside, a wide range of specialized applications can be envisaged in which conventional telephone conversations are enhanced by image transmission. Speech may be encoded at 16 kbit/s, for example, making a substantial part of a 64 or 56 kbit/s channel available for simultaneous still-picture transmission. In the medical case, illustrated in Fig. 1.6(b), the image source could be X-ray plates, CAT scanner output, optical microscope with video camera, and so on. Probably a pointer, driven by a joystick, light-pen or mouse, would also be advantageous.

It can be seen that image communication is not an application in itself, but a service component, along with speech, text, data and so on. From the point of view of the user, the image information component may be very important (e.g. videoconferencing), or even dominant (entertainment TV); it may be less important than another component (e.g. text with illustrations); or the images may be merely icing on the cake (arguably this is the case for the videophone). An indication of the degree of importance is given on a scale 0-10 in the right-hand column of Table 1.3.

Many of the suggested applications are purely speculative at the present time, and it is as well to consider the factors which will influence their chances of success in the market-place.

1.3.2 For and against image transmission

Broadcast television is of course widely available, and a powerful influence in our society, but few of the other image applications have yet achieved significant penetration. To analyse this, the purposes for which images are transmitted must first be considered, and the alternatives to using telecommunications, as tabulated in Table 1.4.

Table 1.4 Purposes of image communication

Category	Purpose	Alternatives
Conversational	See people, their expressions; feeling of presence; impressions	Do without Travel (oneself, or intermediary)
	See material being discussed	Describe in words (spoken) Send photos/video by post, beforehand
Distribution	Entertainment	Travel to cinema, concert, etc Local store, on cassette or disc
	See people and/or discussion material	Do without
	Update many local databases	Send individually, by post or telecomms.
Database	Illustration of text	Do without
	See in order to choose	Enhance textual description to differentiate
	See in order to correlate, identify	Travel to database, or set up local one
Messaging	Desire other party to see images, or respond to other party's request	Send by courier or post
Surveillance	See scene, for security, safety, traffic flow/congestion or other process Identify person or vehicle, etc	Use non-image sensors, or employ a person nearby, or send someone
Processing, info transfer	Images to be processed, compared, used to make new images, or simply for information	Send by courier or post

It can be seen that the advantages arising from the use of telecommunications with images, by comparison with the alternatives, are:

- improved performance — more correct understanding of interpersonal communication and of physical appearance of objects, better appreciation of range of choice to be made, or of security/safety hazard, better ability to comment or ask questions;
- time saving — earlier decision-taking, quicker understanding or appreciation;
- quality of life — avoidance of stress of travel and of the inconvenience of postal transmission, instant availability of entertainment of information and feeling of 'presence', team-building;
- cost saving with respect to viable alternatives;
- projecting a high-tech impression of self or company.

There are, however, a number of factors that weigh against the telecommunications solutions:

- costs of equipment and transmission;
- 'critical mass' — databases need a sufficient pool of users and bodies of provided information; the same for TV distribution such as HDTV; for conversational services there need to be enough relevant locations with compatible equipment;
- image limitations — field of view, clarity (resolution), movement rendition, speed of update (of still pictures), colour accuracy, two- not three-dimensional images,
- human factors — technophobia (difficulty of control, feeling ill at ease in front of a camera, etc.), being unused to facilities (e.g. conference calling), ignorance of possibilities, intrusion into other activity by request for communication.

These factors have different degrees of importance for the different applications — costs are still rather high for many applications, but are falling steadily; technophobia is on the wane; in many cases image limitations are not a problem. Conditions are ripe for an upsurge in the market-place, but there will be a need to tackle first the applications of low critical mass.

It should be noted that some of the listed advantages are met by transmission of still images, individually or in a sequence. Facsimile and SPTV are very economical of transmission by comparison with MPTV, though terminal equipment may be of similar complexity. Facsimile is already widespread; given the limited possibilities for transmitting moving pictures over the general public networks, greater emphasis should be placed on SPTV than is currently the case.

1.3.3 The market-place in 1991

Conversational applications

After a slow start, videoconferencing and teleseminars are enjoying rapid expansion: the world market for such systems reached some $500 m in 1990, exclusive of circuit costs. Though still quite expensive, costs of audiovisual codecs are falling steadily; multipoint systems are of increasing importance as companies expand their conference systems to many sites.

General videophony and audiographic teleconferencing are yet to get off the ground, though there is much developmental activity and numerous trial systems.

The spread of switched digital services, notably ISDN [4], is expected to improve markedly the prospects for conversational applications: it is already possible to create an audiovisual workstation around a personal computer by adding a camera and a codec, and the user is able to make calls in the same way as for the telephone.

It is interesting to note the convergence with telematics, as terminals acquire the full range of communicating functionality, from computer files and electronic mail through facsimile to speech and video.

Distribution

The entertainment market is surging with new business, as satellites and liberalisation bring in more players. Attempts are being made to introduce improved forms of television (see Chapters 2,3), but the commercial difficulties are immense. New markets are emerging in television 'narrowcasting' — for example the distribution of horse race programmes to bookmakers.

Database

There are very few telecommunicating image databases: in fact the success of non-image database services, requiring less expensive terminals, has been so unimpressive that the lack of image databases need come as no surprise. Video library systems have been devised which can be accessed via special local cable-TV networks [5], raising interesting questions on royalties. Some closed-user-group still-picture services have also been reported [6].

Surveillance

The rapid expansion in this market has not yet involved telecommunications to any great extent, though studies [7] showed that this would be an attractive applications area.

Processing and information transfer

True tele-processing applications must probably await the age of intelligent real-time image processing (see Chapter 14), but the transmission of 'unaltered' images is well established — news pictures by wire are an old art, the modern equivalent being electronic news-gathering television. Facsimile has seen an upsurge in the 1980s, but soon signs of saturation will begin to appear, not least because fax is mostly used for alphanumeric information — nowadays this is nearly always created electronically and file transfer may gradually replace facsimile.

1.4. SYSTEMS AND NETWORK ASPECTS

As has been seen, applications can be described in terms of service components or facilities available (audible/visible/tangible) to the users. A telecommunications system is needed because users are remote from each other or from some computer-provided application. This system must provide for transport of information signals between the terminals, and for correct operation of various functions according to the wishes of the user.

The system may be considered in three parts:

- one, two or more terminals appropriate to the user application;
- the network itself providing transport and routing;
- for some applications, specialized equipment located within the network.

1.4.1 Terminals

In general an audiovisual terminal consists of:

- transducers (audio, video, telematic, data);
- signal processing (audio and video coding/decoding, error detection/correction, privacy);
- control (call set-up, end-to-end control and indications);
- multiplexing and network interface.

The structure is considered in more detail in Chapter 7, but certain aspects of the signal processing are worthy of further comment here.

- Audio and image processing are of particular concern, since although transmission costs are reduced by compression, the quality of reproduction may greatly affect the acceptability of the service.
- Transmission errors — audiovisual services are generally more tolerant of transmission errors than, for example, data transmission. Many bits are transmitted but the significance of most of them is quite low, particularly if the signals are not highly compressed. In speech and moving pictures the occasional error may not be sufficiently troublesome to warrant correction. Error correction is necessary for high-quality television transmission (contribution and primary distribution) and

20 AUDIOVISUAL TELECOMMUNICATIONS

also for highly compressed video; in these cases, forward error correction is necessary, since re-transmission would involve too great a delay.

- Privacy — like other digital telecommunications signals, audiovisual services can be encrypted at reasonable cost; there is some demand for privacy in videoconferencing, particularly by satellite where the signal can very easily be picked up by outsiders. There is a particular problem for multipoint conferencing, in that encrypted audio signals cannot be mixed; it is necessary to decrypt at the multipoint control unit, or, if this is unacceptable, then either the audio must be switched (only a single speaker can be heard) or multiple audio channels must be used.

1.4.2 Networks

The properties of telecommunications networks may be classified as follows:

- information capacity of a connection (bit rate, bandwidth);
- geography (coverage, routeability);
- symmetry and configuration;
- performance;
- tariffs.

Capacity and geography

The most geographically universal network is the analogue telephone system, but this is also rather limited in its ability to carry audiovisual signals. Clear still images take many seconds to transmit, and moving images are at present of low resolution and very jerky — see Chapter 5; in neither case is it yet possible to transmit audio simultaneously. For most applications therefore a network with greater information-carrying capacity is sought; at the same time it must be recognized that cost will rise as a function of both capacity and distance, though not linearly in either case.

At the low-capacity end, 64 kbit/s circuits are already becoming common on private networks between digital PBXs, and the ISDN [2] is opening up routeability of 64 kbit/s on a much wider basis. Low multiples of 64 kbit/s are also feasible, with synchronization of the multiple circuits performed at the receiving terminals; however, the implication of the same multiple of call-related tariffs is hard to escape. A feature of the ISDN which should not be overlooked is the extremely fast call establishment — this could be particularly important for services where, for example, image transmission is very intermittent (e.g. browsing through a database).

Higher bit rates are little problem in the case of private or semi-permanent circuits. It is rare that applications require no routeing capability at all, so switching costs (manual, reserved, programmed) remain appreciable. Special-purpose switches are in existence at 2 Mbit/s and higher rates such as 140 Mbit/s, and video switches for analogue television.

Automatic (user-controlled) switching at higher rates than ISDN are contemplated for the future, but the investments will have to be very great. For this reason the asynchronous transmission mode (ATM) solution is finding considerable favour, because a single switch can handle all bit rates up to a certain maximum: around 150 Mbit/s. Such a concept is very interesting for audiovisual communications, because it offers the potential for considerable flexibility in the choice of quality of (moving) video signal. The rate could be chosen on a call-by-call basis, according to the quality desired; there is even the possibility of varying the bit rate according to the changing information in the moving image (higher bit rate during movement or scene change — see Chapter 9), though the advantages of such an approach in transmission cost terms have yet to be proved.

What is the 'ultimate' bit rate for (moving-picture) audiovisual services? There is no easy answer, for while image-compression techniques have advanced greatly, the basic costs of bit transport are also rapidly falling. The balance of probability is that bit rates higher than those of basic ISDN will gradually prevail, for the following reason.

Subjective performance at 64 and 128 kbit/s is well below that of entertainment television, which the general populace has come to expect (a quality which itself will be improved in the next decade); even though picture rendition may be adequate for many conversational purposes, the codecs introduce a delay which is rather worse than that encountered on satellite circuits — a considerable detriment to free discourse. The application of model-based processing techniques, as in Chapter 11, might reduce the delay to more modest values, but this will not be possible at acceptable cost for at least five years.

In the early days audiovisual services are most likely to flourish in the business environment, particularly in 'nearly-closed' user groups; it is in this area that bit rates will be least constrained by the cost factor — not only are businesses more prepared to pay for higher performance, but the competitive provision of virtual private networks will ensure that bit rates become of decreasing significance in the cost equation. Rapid growth in local-area networks of high bit rate can be expected during the next decade, these carrying audiovisual services at rates in the range 384 kbit/s to 2 Mbit/s; the LAN islands will be interconnected with high-capacity links in a fully managed network, initially on an intra-company basis, with gateways to the public switched digital network at 64 kbit/s and low multiples thereof. As shown in Chapter 7, the standardization of the audiovisual signal infrastructure has been conceived with just such a freedom of choice of bit-rate in mind.

Symmetry and configuration

Of the applications listed in Table 1.3, the conversational services are generally symmetrically bidirectional, while the remainder are not. It is true that human conversations are almost entirely half-duplex (use is made of this property for multiplexing on some long-distance circuits), but in videophone and videoconferencing the image must be transmitted even when a person is not speaking. Teleseminars, training sessions and the like tend to be asymmetric, especially when those addressed are at several locations.

Surveillance, database, and messaging services are highly asymmetric, particularly where large volumes of image information are involved; there is generally a relatively small flow of information in the reverse direction for control purposes. Only in the case of input to a database is substantial transmission required in both directions.

Most switched telecommunications networks, including the new 'basic' ISDN, provide only symmetrical connections (the exception is the contribution and distribution of entertainment television, provided on special high-capacity networks). However, the future broadband ISDN, using ATM, is not inherently symmetrical — transmission in the two directions is independent and variable: this will make possible the efficient carriage of all types of service.

Virtually all switched calls are point-to-point, though modern PABXs have provision for bridging a number of speech calls. For audiovisual teleconferencing it is necessary that the network provide call-control capability to connect a number of terminals to a multipoint control unit (MCU), and even between two or more MCUs.

Performance

The small amounts of jitter and errors normally occurring in digital networks are readily handled within the audiovisual systems in such a way that the user is unaware of them. Furthermore, following an error burst, slip, or temporary break the systems will quickly reset, minimizing the interruption of the application.

Tariff strategy

Digital systems are essentially bit transport, but to charge simply on the basis of each bit transported is by no means the most satisfactory method. From the user's point of view, the value added per bit of financial data, for example, is very high; the value per bit of transported speech is very much lower, and that of vision lower still. A transaction may involve only a few thousand bits, while at the other end of the scale a videoconference call might involve over a gigabit. A further factor is that distance is often irrelevant to the user, except in travel-substitution applications; at higher bit rates there is a significant distance-dependent element of cost, while at lower bit rates the cost tends to be dominated by switching and administration; this tendency is increasing as the costs of bit transport on trunk routes falls, with the introduction of high-bit-rate fibres.

To optimize the economies of audiovisual services, from the point of view of both user and network provider, the tariff strategy could be different from that for services of high bit-value, using one of the following approaches.

- Service class — distinction is imposed in a regulatory fashion, in which the terminal transmits an indicator to the network, and the latter then charges for calls at a rate approved for the service.

- Distinction by call characteristics:

  — charges are made both for call set-up and duration, encouraging the use of multiple short calls for rapid image retrieval, without the burden of keeping open a high-speed channel during the much longer periods while the images are being inspected; this approach would also be beneficial in security surveillance applications, the transmitter sending intermittent 'confidence' images for the long periods when there is no scene change.

  — charging rate declines with call-holding time (effectively a form of bulk discount), encouraging the use of teleconferencing systems.

  — charging rate is not proportional to bit rate (for example, 384 kbit/s charged at 2.5 times that for 64 kbit/s), encouraging a trend to better picture quality and higher bit rates.

In the latter respect, the proposed ATM networks offer a somewhat better prospect, provided that they are designed from the outset with the appropriate tariff structure in mind. Here again a simple price per bit would be wholly inadequate. By making a distinction between audiovisual and other types of information, and hence charging for this on a different scheme, it would be possible to offer an audiovisual transport service at one of a range of qualities to suit market demands, all of these being handled by the single transport system. It would still be possible to make a charge for each call set up, but other than this it would be difficult to introduce dependency on call-holding time. In the busy hour, rather than having to introduce heightened tariffs, the quality might simply be reduced at the same tariff by reducing the throughput as the circuit becomes heavily loaded.

In conclusion, there is a real need to consider the influence of tariffs on the potential takeup of audiovisual services, which in many cases have different characteristics from the well-established telephone services.

1.4.3 Specialized network equipment

Some equipment providing special facilities for audiovisual services appears from the user's viewpoint to be located within the network. However, from the network point of view such equipment is merely another terminal, with multiple ports and automatic call-handling capability.

Multipoint control for conversational services

It is sometimes desirable to connect three or more terminals together, such that all parties may hear each other and receive as much additional visual information as practicable. This facility is becoming increasingly important in formal teleconferencing, which often involves people from more than two locations.

It may also be useful to establish conference calls on an *ad hoc* basis. Admittedly this facility is little used in current telephony, but this may well be because the poor human factors of call set-up militate against adoption by users. On the other hand it could be argued that, if call set-up were sufficiently user-friendly, conference calls might be a better reason for the adoption of videophone than mere enhancement of telephony.

Multipoint control units are discussed further in Chapter 6. It will be important that such units be able to cope with a range of terminals that are not identical — the adoption of a range of standardized but interworkable infrastructure signals will be particularly valuable in this respect.

Databases

At the present time most databases provide information in text form, with a limited amount of graphics. Audio will become important both for information output and also for control input. Image databases, both moving and still (video and facsimile formats) will become increasingly important as storage costs fall.

1.5. THE FUTURE

There has been a great upsurge in the applications of video over the past decade; people are much more aware of the possible applications of video, and are less afraid of the TV camera. Social barriers to the introduction of visual telecommunications services are therefore being lowered, and at the same time the pressures are increasing for the better use of technology in business. Despite the frequent optimistic press reports, developers of visual telecommunications have always been well aware of the problems of getting business people to want such services.

Audiovisual communication is very often not in itself a solution to the problems of business information technology. For example, though image communication could considerably enhance the performance of medical services, there are many other aspects of databases, communications and computing which are probably more important. Again, teleworking may be seen as increasing towards the 21st century, and audiovisual communications an essential ingredient of the teleworker's links; more important, however, will be the successful integration of the personal computer into the electronic office environment (not necessarily paperless), the security aspects, and the improvement of the human factors. For the office worker who is able to operate from home, the videophone may be increasingly important as a means of maintaining better relationships with co-workers (coffee break by video?), but transmission costs would have to be kept very low. As an example of teleworking, take the case of a telephone answering service staffed by young mothers, the busy hours occurring while the children are in school; socially, it would be good to have a full-time open link to a multipoint unit, so that the mother can be continuously aware of the others in her group, and can chat at slack moments.

It is certain that by the year 2000, the volume of business audiovisual communications will have increased five- or ten-fold from its present level. At this time it is important to engage in practical trials of potential applications, to clarify the terminal and network requirements and to make potential users aware of the possibilities.

However, the other factors listed in section 1.3.2 must be considered, both for and against the applications. To make further improvements to moving picture quality at low bit rates, improvements in chip technology alone will not be enough — new algorithms will be needed, almost certainly involving model-based image processing.

Current capabilities in audiovisual telecommunications result from rapid technological progress over the last decade, particularly in respect of TV cameras and displays, RAM, DSP, and ASIC capabilities, together with advances in image-compression methods. It is reasonably certain that these trends will continue, which is just as well — since costs for many applications are still too high.

Advances in the soft display of documents will be needed — the limitation of present TV and PC screens is severe, and HDTV is probably not the answer. As to the quality of picture (non-document) images, while it is not yet certain that HDTV will become widespread there will be a gradual improvement in the quality of domestic entertainment TV and this will make the use of lower-quality images for videoconferencing and videophony less acceptable in the long run.

By comparison with speech, text and document transmission, visual telecommunication is in its infancy; but the use of images, still and moving, is growing rapidly in all walks of life, and it can be predicted with great confidence that the electronic communication of these images will grow rapidly in the 1990s to become a very significant part of the total information flow.

REFERENCES

1. CCIR Recommendation 601: 'Encoding parameters of digital television for studios', ITU, Geneva (1990).

2. Gonzalez R C and Wintz P: 'Digital Image Processing', Addison Wesley (1987).

3. CCITT Recommendation I.211, ITU, Geneva (1988).

4. Griffiths J M, Ed.: 'ISDN Explained', Wiley (1990).

5. Kerr G W: 'On-demand interactive video services on the British Telecom switched-star cable-TV network', Br Telecom Technol J, 4, No 4 (October 1986).

6. 'Visual communication services for VI&P', Special feature (1) in NTT Review, 3, No 5 (September 1991).

7. Reid G M et al: 'Application study for still-picture television', Br Telecom Technol J, 3, No 3, pp 62-72 (July 1985).

2

BROADCAST TV SYSTEMS

S Searby and P M Elmer

2.1. INTRODUCTION

This chapter outlines the current broadcast chain and how new developments in TV may influence it. The standards which define the television signal are discussed first and the concept of high-definition television (HDTV) is introduced. HDTV standardization activities are described, with particular reference to the work of the International Radio Consultative Committee (CCIR). Various ways of transmitting the television signal over analogue and digital circuits are covered before the structure of the current broadcast chain is described. Finally the chapter explains some of the developments which might influence the delivery of television services in the future.

2.2 STANDARDS FOR TELEVISION AND HIGH-DEFINITION TELEVISION

2.2.1 The evolution of television and its problems

Today's TV systems have evolved from the first monochrome versions of the 1930s. In selecting the fundamental system parameters, such as number of lines/picture, number of frames/second and picture aspect ratio (ratio of width to height), constraints were imposed by the characteristic bandwidth limitation and distortion of AM radio channels. Further technological limitations of the camera and display had also to be considered. The monochrome system was designed to employ the same video signal specification from studio to receiver: the problems of such signal uniformity will be discussed later.

The first key instance of the need for compatible evolution (although at the time this could not be achieved) was the upgrading of the 405-line system to 625 lines in the 1950s. During the change-over period, a considerable investment was necessary in the form of new studio equipment, simultaneous radio transmission of both standards and development of dual-standard receivers.

The 625-line system provided a good level of picture quality when viewed from a distance of 6 times the picture height (6H). Only 576 lines carried picture information, the remainder being used in synchronization and scanning. Twenty-five pictures per second would be sufficient for portrayal of movement but the viewer would then be subjected to a considerable amount of 'large area flicker'. To reduce this flicker to an acceptable level the display needed to be refreshed at twice this rate — doubling the bandwidth of the signal if all the lines were reproduced in each picture. It was decided that interlaced scanning provided the best compromise as it doubled the refresh rate whilst occupying the same bandwidth. This was achieved by dividing the picture into two interleaved sets of lines (fields): each field could then be displayed in half the picture period. Such a technique could not be applied without penalty. It introduced 'inter-line flicker', most noticeable on high contrast horizontal edges, and blurring of vertical edges during periods of rapid motion. Another problem of this standard resulted from the need to view the screen from 6H to avoid line structure. At this distance the viewer lacked involvement with the programme material as it appeared as a 'picture in a box'. Advanced TV systems will concentrate on elimination of the above problems of 625-line television.

Appreciating now the need for compatibility, considerable effort was made to find a means of upgrading the monochrome system to colour in a compatible manner. A frequency-division multiplex technique was employed, wherein colour signals were modulated on to a sub-carrier whose phase changed on alternate lines (PAL) [1]. By exploiting the colour disparity characteristics of the human visual system, colour-difference signals could be generated and bandwidth limited to as little as a quarter of that of the monochrome (luminance) signal without any degradation in observed picture quality. Luminance and colour-difference signals were derived from the red, green and blue (RGB) primaries of the colour camera:

luminance $\quad E_Y = 0.30\, E_R + 0.59\, E_G + 0.11\, E_B$

colour-difference $\quad E_{Cr} = S_{Cr}\, (E_R - E_Y)$
$\phantom{\text{colour-difference} \quad} E_{Cb} = S_{Cb}\, (E_B - E_Y)$

Scaling factors (S_{Cr} and S_{Cb}) applied to the E_C signals served to match the colour-difference amplitude to that of E_Y. The PAL composite signal suffered from artefacts such as cross-colour, caused by high-frequency luminance information being decoded as low-frequency colour. (The effect is often noticed as coloured striping in fine details of a checked jacket.) Since the amplitude and phase of the colour subcarrier needed to be maintained precisely it was necessary to define stringent specifications for the characteristics of analogue transmission channels. Such specifications were derived in two stages: subjective testing methods were used for assessing picture quality [2] and then these techniques were applied to define a relationship between the observed (subjective) picture quality and objective measures such as channel noise [3].

Compatibility of the PAL signal with the old monochrome signal was to prove particularly beneficial to the studio, network and viewer. Existing monochrome receivers could accommodate the new colour signal albeit with minor picture artefacts such as subcarrier crawl; while colour receivers could accept monochrome transmission of archived material.

2.2.2 The move towards digital television

The problems of operating with PAL signals, taken in conjunction with the technological developments which made digital processing possible, encouraged a third major change in television. The advantages of digital processing in the studio were the reproducibility and flexibility in recording and special graphic effects. The analogue component signals E_Y, E_{Cb} and E_{Cr} were used as a three-signal set for equipment interconnection. This important change was an essential part of the decoupling of the studio signal standard from that of transmission. Within the studio, signals were handled in analogue or digital component form, and PAL-encoded just prior to transmission. The studio benefited from an increased tolerance of channel imperfections and much improved colour signal quality.

An all-digital specification was published by the CCIR in 1986 [4]. Recommendations 601 and 656 specified the encoding parameters of digital television for studios, and interfaces for digital component video signals in 525- and 625-line television systems. Worldwide adoption of these recommendations in systems using 50 Hz and 60 Hz field rates saw the beginning of the digital television era. The major elements of Recommendation 601 are detailed in Table 2.1.

30 BROADCAST TV SYSTEMS

Table 2.1 Digital encoding of 625-line television to Recommendation 601

Parameter	Value	
coded analogue signals	E_Y	E_{Cb} and E_{Cr}
total samples per line	864	432
active pixels per line	720	360
sample structure	orthogonal in space and time	
sampling frequency	13.5 MHz	6.75 MHz
digital coding	8 bits/sample	
quantization levels	220 levels black = 16 white = 235	225 levels min = 16 max = 240
synchronization	levels 0 and 255 reserved	

The total bit rate of signals conforming to Recommendation 601 can now be calculated as the product of pictures per second, total lines per picture, samples per line, and bits per sample for each component. The resulting bit rate is 216 Mbit/s. A similar calculation for the active picture results in a value of only 166 Mbit/s, which is 76% of the bit rate for the total picture. This highlights the overhead in the analogue format for synchronization and other operations of the scanning process. By substituting digital synchronization words, more efficient digital transmission can be achieved. Further reduction in bit rate can be made by removing psychovisual redundancy using various picture coding methods.

At the present time, many studios employ digital signals in equipments such as vision mixers, switchers and recorders using the so-called 4:2:2 digital interface, named after the ratios of the luminance and colour-difference sampling frequencies.

2.2.3 High-definition television

Studies to find a successor to conventional television began in the 1970s with the ultimate aim of providing an increased sensation of reality to the viewing environment. A fundamental conclusion of these studies was that an increase in picture width provided a better match to the field of view of the human visual system, as did an increase in total picture area [5]. These changes could be made by increasing the aspect ratio and screen size of conventional television, but in so doing the viewer would perceive a reduction in resolution, and increased line visibility. To reduce these effects a new scanning format was needed to generate more TV lines, and an increase in bandwidth was then necessary to accommodate the higher horizontal resolution and wider picture format. Another factor affecting the realism of television was the non-ideal chromaticity of the system. Modification to the spectral location of colour primaries could usefully extend the colour gamut, especially in flesh-tone reproduction. The conclusions of this work recommended a viewing distance of 3H, a viewing angle of 30°, 1222 lines per picture and a minimum colour primary bandwidth of 19 MHz [6].

In 1974, the CCIR decided that the following question should be studied: 'What standards should be recommended for high-definition television systems intended for broadcasting to the general public?' This was justified on the basis of the need for improved picture quality, the likely availability of wideband communication systems and the progress in development of large screen displays. From this point on, the study of HDTV was to increase and the political and economic implications were to become apparent.

Proposals for a world production standard for HDTV have been submitted to the CCIR [7]. The two main proposals, originating from Europe and Japan, are related to the different field frequencies of 50 Hz and 60 Hz, and issues regarding compatibility with existing television standards. The USA is concentrating, at present, on the selection of a terrestrial broadcast standard for enhanced television and HDTV. Whatever scheme it chooses it will clearly wish to have a production standard which is closely compatible.

The European and Japanese scanning standards are defined in Table 2.2.

Table 2.2   Scanning specifications for HDTV

Parameter	European value	Japanese value
lines per picture	1250	1125
active lines per picture	1152	1035
pictures per second	50	60
interlace ratio	1:1	2:1
aspect ratio	16:9	16:9
primary colour bandwidth	60 MHz	30 MHz

A new aspect ratio of 16:9 is 33% higher than before, and achieves a 35° viewing angle when viewed from 3H. In the European case 1250 lines achieve a doubling of vertical resolution capability, and the increased bandwidth gives twice the spatial resolution on the wider picture. By doubling the number of lines of the 625-line system exactly, compatibility between the new and old systems is easily achieved by 'simple' processing. Interlace artefacts are removed by the adoption of progressive scanning (interlace ratio 1:1), which further increases the perceived vertical resolution of pictures, and the picture rate. At 50 Hz, motion rendition is adequate, but large-area flicker can still be annoying. By further exploiting the decoupling between transmission and display standards, field rate upconversion [8] can be performed at the display to achieve a 100 Hz refresh rate.

In the digital domain, the parameters are developed from the CCIR Recommendation 601, as shown in Table 2.3. The value of E_Y is generated

32 BROADCAST TV SYSTEMS

Table 2.3 Digital encoding of 1250-line HDTV

Parameter	Value	
coded analogue signals	E_Y	E_{Cb} and E_{Cr}
total samples per line	2304	1152
active pixels per line	1920	960
sample structure	orthogonal, picture repetitive	line quincunx, picture repetitive
sampling frequency	144 MHz	72 MHz
digital coding	8 bits/sample minimum	
quantization levels	under study	
synchronization	under study	

by a new linear equation in R, G and B, which are themselves derived from a new set of chromaticity co-ordinates to improve colour rendition. A line quincunxial sample structure is used for colour-difference signals. This matches the resolution of the colour signals to the spatial response of the human visual system, which is deficient in the diagonal directions, by a sample interlacing technique which halves the sampling rate. This highest-quality signal is referred to as high-definition progressive (HDP) and heads a proposed hierarchy of HDTV signals [9].

The digital encoding parameters of Table 2.3 present a significant technological challenge to the system designer, but result in an excellent picture quality. Table 2.4 shows a hierarchy of formats, which can be derived from HDP, along with their data transmission requirements.

Table 2.4 HDTV hierarchy with bit-rate requirements

Format	Luminance sampling frequency (MHz)	Colour-difference sampling frequency (MHz)	Total bit rate (Mbit/s)	Notes
HDP	144	72	2304	
HDQ	72	36	1152	As HDP but with quincunx sub-sampling for luminance and orthogonal sub-sampling for colour difference
HDI	72	36	1152	Interlaced format with orthogonal sample structure
HDI	54	27	864	Reduced sample rate — used as input to HD-MAC encoder

A picture coding technique developed as part of the Eureka-95 initiative has a compression factor of 2:1 and allows transmission of HDI on standard 560 Mbit/s digital networks [10,11]. Transmission at 140 Mbit/s is possible with good picture quality by more complex transform-based coding methods. A further economy can be achieved by reducing the HDI sampling frequencies by 25% to 54 MHz and 27 MHz for E_Y and E_C. This signal contains twice the number of samples of Recommendation 601 and is used as the input to the HD-MAC bandwidth reduction encoder for HDTV broadcasting. Further consideration is given to the performance aspects of HDTV in Chapter 3.

2.3. DELIVERY SYSTEMS

The previous section described picture standards for normal and high-definition television and introduced the idea of luminance and colour-difference component signals. In order to convey these signals from a source (normally a camera) to a display in a television receiver a delivery system is required. This will involve coding the component signals in some way and transmitting them over a suitable medium. Various ways of performing coding, modulation and transmission are explained below in order to provide background information before describing the TV broadcast chain. The section goes on to describe an HDTV broadcast chain and finally looks at possible future transmission of broadcast services on broadband networks.

2.3.1 Coding, modulation and transmission

Figure 2.1 shows the various alternative processes which will normally be applied to the component signals before they are suitable for connection to the transmission medium. The same processes will be applied, in reverse, at the receiver. The processes are separated into three stages of which one, or both, of the first two may be omitted in different circumstances. The combinations of these processes will be discussed later when the broadcast chain is described.

- The first stage can be used to merge the separate components into a single combined analogue signal. The schemes normally used for this are shown in Fig. 2.1 for frequency domain and time domain techniques.

- The second processing stage is used if the signal is to be conveyed digitally. Probably the best known is PCM (pulse code modulation) [12,13] but PDM (pulse density modulation) [14] could be employed to allow fairly simple and cheap decoding. More sophisticated schemes such as DPCM and transform coding may follow the digital sampling in order to reduce the transmission bit rate.

34 BROADCAST TV SYSTEMS

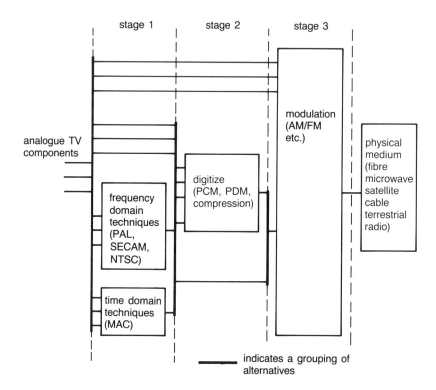

Fig. 2.1 Alternative processes to code TV signals for transmission.

- Before the signal, now either analogue or digital (effectively analogue but having discrete values), is applied to the physical transmission medium it will normally need to modulate a carrier signal. This corresponds to the third stage in Fig. 2.1 and will be achieved by amplitude or frequency modulation or a variant of one or other. If several channels are to be carried then separate carriers can be used. In the case of digital signals, a time-division multiplex may be used before modulating the carrier.

In the case of HDTV the same stages will be used as shown in Fig. 2.2. However, the schemes now proposed for combining components tend to be either hybrid time domain/digital or frequency domain/digital schemes. This means that part of the picture is coded in analogue form but supplementary information is transmitted digitally and is interpreted at the receiver to enhance the picture.

DELIVERY SYSTEMS 35

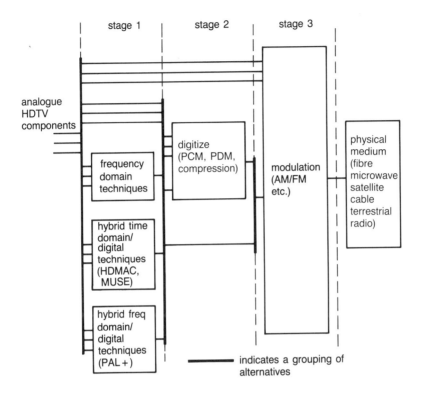

Fig. 2.2 Alternative processes to code HDTV signals for transmission.

2.3.2 The existing TV broadcast chain

The diagram in Fig. 2.3 gives an outline of the structure of the current broadcast chain. The links have been broadly categorized into contribution, primary distribution and secondary distribution.

The contribution links require the highest quality and are used where further processing of the video is expected (e.g. chroma-key which is highly sensitive to picture impairments) before onward transmission. The majority of these links use terrestrial microwave radio. Coaxial cable is used in some cases and standards exist for TV transmission on 12 MHz, 18 MHz and 60 MHz carriers. International connections can be provided using terrestrial microwave or satellite circuits. A variety of connections can be used for provision of outside broadcasts — over telephone pairs for short distances, over coaxial or polyquad cable, or by small dish microwave systems operating around 7 or 12 GHz.

36   BROADCAST TV SYSTEMS

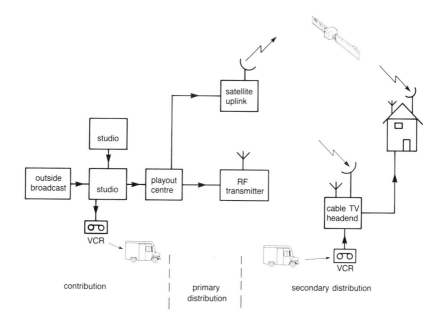

Fig. 2.3   The present TV broadcast chain.

The primary distribution links carry signals which do not require further processing and again analogue transmission over microwave relay and co-axial cable dominates. Secondary distribution provides delivery of signals to the home. The main systems in use are cable distribution in the VHF band, terrestrial radio in the VHF and UHF bands and satellite broadcasting.

Almost all of the above links carry composite analogue signals; separate video components are only carried on a small minority of contribution circuits.

Digital transmission using optical fibre (and sometimes microwave radio) is now being introduced for contribution and primary distribution. The coding used for the various standards and bit rates is described in CCIR Report 1089 [15].

2.3.3   A high definition TV broadcast chain

For the introduction of HDTV broadcasting a number of enhancements to the broadcast chain could be made. Figure 2.4 shows a typical HDTV chain.

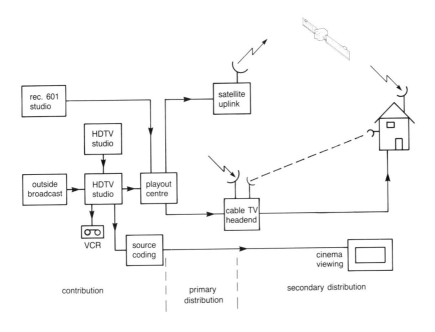

Fig. 2.4 An HDTV broadcast chain.

Contribution links would carry component signals in a digital form. Suitable schemes for compression include variable-site sub-sampling [10,11] and DCT [16]. A number of contribution codecs were developed in Europe for transmission of HDTV at 140 Mbit/s (e.g. in the RACE Hivits project R1018 [17]).

Various hybrid schemes exist for distribution of HDTV: HD-MAC [18] and MUSE [19] for satellite distribution and PALplus [20] (an enhanced TV format) for terrestrial radio. The same schemes are also suitable for distribution over cable. Primary distribution from the playout centre to satellite uplink or cable headend would probably use a digitized version of the secondary distribution signal. In the USA a number of Advanced TV proposals for terrestrial broadcasting are under evaluation, of which half are entirely digital. The International Standards Organisation is defining standards for digital coding of TV pictures at rates up to 10 Mbit/s and is expected to progress from this to coding of HDTV (see Chapter 10). Such schemes provide the possibility for the introduction of an entirely digital broadcast chain. It is also possible to code the pictures in such a way as to allow lower-resolution pictures to be extracted easily. This is termed 'compatible' coding and is currently the subject of study in RACE.

2.3.4 Future broadcast scenarios using optical fibre distribution

The transmission chains described so far have been made up of semi-permanent circuits with distribution to the home using traditional methods such as terrestrial radio. Within the CCITT and RACE there is considerable effort being directed towards defining a B-ISDN. If such a network were to be installed then it would be capable of carrying TV and HDTV (along with many other services) direct from a studio or playout centre to a customer's home. A number of alternative scenarios may exist as evolutionary steps on the route to a fully integrated network. Such networks would provide optical fibre into the customer's home with TV and HDTV distribution sharing the fibre with the B-ISDN circuit.

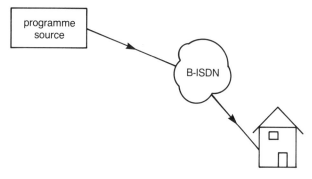

Fig. 2.5(a)   TV to the home using B-ISDN.

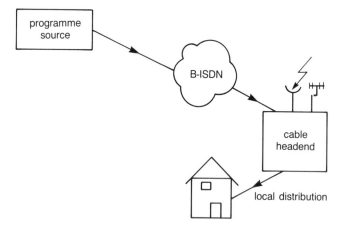

Fig. 2.5(b)   TV to the home using local distribution.

In Fig. 2.5(a) the target B-ISDN situation is compared with an alternative local distribution approach in Fig. 2.5(b). The cable headend is able to receive programme material from conventional off-air broadcasts as well as taking inputs via the B-ISDN. The use of separate local distribution takes advantage of the fact that the TV channels are being shared among many customers and so makes wider bandwidths available.

There are three basic architectures to consider for this local distribution. These are the 'big star', 'double star', and passive bus or 'tree and branch'. These architectures are described below with particular emphasis on the passive bus which offers some important advantages over the other two. Ring-type architectures are probably more appropriate to customer premises networks and so are not discussed here.

The 'big star' architecture would provide a unique dedicated link from each customer to the central exchange (see Fig. 2.6). Each link could consist of one or more fibres, depending on the capability of the terminal equipment at each end. This architecture has two major problems. Firstly, it is very expensive in terms of fibre and optoelectronic devices and therefore is cost-effective only for large business customers. Secondly, if deployed in large quantities it would result in many fibres originating from the central exchange. This would require large fibre cables, with hundreds or even thousands of fibres per cable. This could result in very high repair costs in the event of accidental damage.

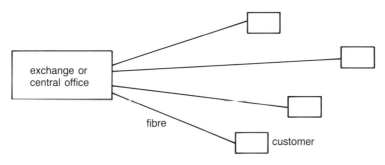

Fig. 2.6   'Big star' architecture.

The 'double star' architecture requires fewer fibres at the central exchange, but uses an electronic switch in the field [21], as shown in Fig. 2.7. This has the disadvantage of requiring sophisticated electronics in an external environment, with associated cost, security and maintenance problems. In addition, the switch design would probably impose a practical limitation on the services or bandwidth it could carry. All switches would then require modifying or upgrading when a future service was introduced which required greater bandwidth or a different transmission method.

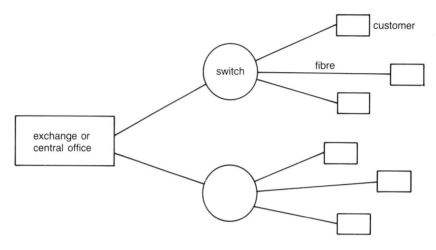

Fig. 2.7   'Double star' architecture.

The passive bus, or tree and branch architecture, has none of these disadvantages. It is economical, as exchange equipment and fibres are shared between many customers, and has no active electronics in the external environment (see Fig. 2.8). Furthermore, as the network is purely passive, it imposes no restrictions on bandwidth, wavelength or transmission methods. The terminal equipment will need to be designed and upgraded accordingly, of course, but the fibre infrastructure will be 'future-proof'.

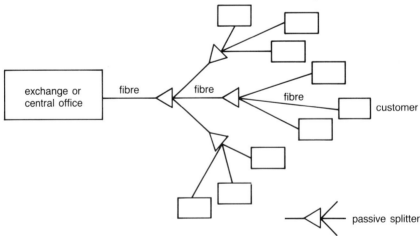

Fig. 2.8   'Passive bus' architecture (passive optical network).

DELIVERY SYSTEMS 41

The passive optical network (PON) [22] comprises single mode fibre and a number of passive optical splitters. The size and location of the splitters depends on local topography, the transmission bit rate and the optical power budget. As the fibres are shared by many customers, PONs require a transmission method that takes this into account. There are three basic multiplexing techniques that could be used alone or in combination. They are time, frequency and wavelength division multiplexing (TDM, FDM and WDM). In Fig. 2.9 TDM is used for the telephony/ISDN channel from the exchange to the customer, with each customer allocated a specific number of time slots, depending on the services required. In the upstream direction, the customer's terminal is programmed to transmit only in its allocated time slots, thus ensuring bit synchronization back at the exchange. A second wavelength provides a number of distributed TV channels. The TV channels are multiplexed electrically, using FDM techniques, and the resultant electrical signal is used to modulate the output power of the optical transmitter. This technique is known as subcarrier multiplexing [23] and the decoding of the TV channels can be achieved using a satellite-type receiver. Alternatively, the TV channels could be digitized, and multiplexed using TDM, as shown in Fig. 2.10. As the TV is carried as a digital signal a codec will be required, but as the bit rate allocated to each channel can be high the use of compression coding may not be necessary. A simple, low-cost scheme such as PDM (mentioned in section 2.3.1) is well suited for this application and offers transmission which is transparent to the TV or HDTV format being used on the various channels.

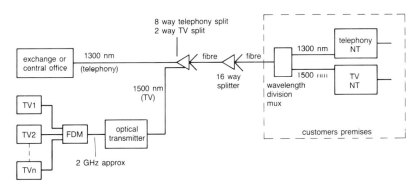

Fig. 2.9 Telephony and TV on PON, using WDM, FDM and TDM.

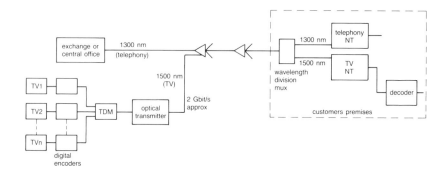

Fig. 2.10 Telephony and TV on PON, using WDM and TDM.

RACE project R1051 concentrated on networks similar to that in Fig. 2.10. With TV distribution, bit rates are around 10 Gbit/s. R1010 studied coherent multichannel techniques to allow a large number of closely spaced wavelengths to be used.

2.3.4 Future non-fibre distribution

In some circumstance non-fibre distribution is appropriate. RACE Mobile project R1043 is mainly concerned with consideration of a universal mobile telecommunications system (UMTS) which will provide a unified alternative, with extended coverage, to the disparate mobile systems presently evolving. The aim is a low-cost system available to a very high proportion of the general population, for business and private use, everywhere.

However, operating in the radio frequency region of 2 GHz with an expected bandwidth assignment of about 200 MHz, the user data rate provided is likely to be limited to basic ISDN services. A minimum capability of 2B + D access is certain but still under examination is the possibility of extension to 2 Mbit/s with some degradation in integrity. Consequently, part of R1043 is concerned with the problems of extending broadband ISDN services and other integrated broadband communication services, into the mobile sector, with data rates from 2 Mbit/s to over 70 Mbit/s. Such a system, called mobile broadband system, needs a minimum of 2 GHz bandwidth of available radio spectrum to support an economic number of users who require TV transmission facilities. This amount of radio spectrum is likely to be available only at millimetre wavebands. At these frequencies propagation is limited to line-of-sight which implies small microcell sizes.

2.4. CONCLUSIONS

This chapter has discussed the standards defining TV and HDTV signals as a background to a description of various broadcast systems. It has described the operation of the current broadcast chain followed by a higher-definition broadcast chain, which may come into operation in the next few years, and has introduced, finally, some possible scenarios for the longer term.

At the present time there is considerable speculation as to how and when high-definition broadcasting may be introduced. It seems likely that in the longer term broadcasting will be done digitally, allowing, in addition to the usual advantages of digital transmission, a greater number of channels to be provided to the customer. New distribution formats, however, take many years to become established and commercial factors normally determine whether they will succeed.

REFERENCES

1. 'Specification of standards for 625-line system I transmissions in the UK', DTI Radio Regulatory division, London (1984).

2. Allnatt J W: 'Transmitted picture assessment', John Wiley and Sons Ltd (1983).

3. Macdiarmid I F and Allnatt J W: 'Performance requirements for the transmission of the PAL coded signal', Post Office Telecommunications Research Department Report 691 (1978).

4. 'Recommendations and reports of the CCIR', XI , Part 1, Geneva (1986).

5. Hatada T, Sakata H and Kusaka H: 'Psychophysical analysis of the sensation of reality induced by a visual wide field display', SMPTE Journal, 89 (August 1980).

6. Fujio T: 'A study of high-definition television system in the future', IEEE transactions on broadcasting, BC-24 , No. 4 (December 1978).

7. 'Conclusions of the extraordinary meeting of study group 11 on high-definition television', Geneva (1989).

8. Childs I and Roberts A: 'The compatibility of HDTV sources and displays with present day television systems', Journal of the IERE, 55 , No 10 (October 1985).

9. Nasse D and Chatel J: 'Towards a World studio standard for high-definition television', SMPTE Journal (June 1989).

10. Elmer P M: 'Variable compression coding of HDTV signals', Proceedings of the second international workshop on signal processing of HDTV, L'Aquila (29 Feb—2 Mar 1988).

11. Elmer P M: 'The design of a high bit rate HDTV codec', Proceedings of the third international workshop on HDTV, Turin (30 Aug—1 Sep 1989).

12. Cattermole K W: 'Principles of pulse code modulation', Iliffe (1969).

13. Devereux V G and Jones A H: 'Pulse code modulation of video signals: codes specifically designed for PAL', Proceedings of the IEE, 125, No. 6, pp 591—598 (1978).

14. Burgess G D: 'A pulse density modulation video codec', BT Technol J, 9, No 2 (April 1991).

15. CCIR Report 1089 (MOD F), 'Bit-rate reduction for digital television signals' (January 1990).

16. Barbero M and Stroppiana M: 'Digital coding of HDTV based on discrete cosine transform', ITU-COM 89 Symposium, pp 235—239, Geneva (1989).

17. 'RACE '90 — Research and Development in Advanced Communications Technologies in Europe' (March 1990).

18. Annegarn M J J C et al: 'HD-MAC: a step forward in the evolution of television technology', Philips Technical Review, 43, No 8 (1987).

19. Ninomiya Y, Ohtsuka Y and Izumi Y: 'A single channel HDTV broadcast system — the Muse —', NHK Laboratories Note, No 304 (September 1984).

20. Messershmid Z: 'Palplus — an approach to meet the challenges of the future for existing PAL services', IBC90, pp 58—59, Brighton (September 1990).

21. Fox J R and Boswell E J: 'Star-structured optical local networks', Br Telecom Technol J, 7, No 2 (April 1989).

22. Taylor C G and Fox J R: 'Television distribution over passive optical networks', International TV Symposium, Montreux (June 1989).

23. Rosher P A et al: 'Broadband video distribution over passive optical networks using subcarrier multiplexing techniques', Electronics Letters, 25, No 2, pp 115—117 (January 1989).

3

TELEVISION OF IMPROVED QUALITY

B R Patel

3.1. PROSPECTS FOR HIGH DEFINITION TELEVISION BROADCASTING

The availability of narrow-band satellite channels is seen as an excellent opportunity for the introduction of wide-screen high-definition (HD) television services. In Europe, multinational coverage can be achieved in the BSS (broadcast satellite systems) band using a high-power DBS (direct broadcasting by satellite) and 27 MHz channel bandwidth. This coverage can also be supported in the FSS (fixed satellite systems) band, operating at about 10 dB lower power and a larger channel bandwidth. Although these channels were originally planned for conventional definition (CD) television using 45 cm aerials, it has been shown that an adequate HD performance is possible with 45 and 90 cm aerials. Aerial sizes of 1.4 times these values, however, are recommended for noise performance exceeding that of existing composite television. A 50 Hz HD system for satellite use has been demonstrated to be also suitable for cable transmission, but requires a 50% additional bandwidth over the 8 MHz per channel presently used for composite television.

The 50 Hz HD system is compatible with the new MAC transmission format proposed for CD television in Europe, such that either can be received on both CD and HD receivers. A combination of digital and analogue techniques is used for its bandwidth compression. In non-evolutionary approaches to HD television such as MUSE [1] and all-digital systems (presently in the process of evaluation for North America, primarily for delivery over terrestrial media), simultaneous broadcast (simulcast) will be required to serve existing CD television viewers. In order to allow early commitment to HDTV, with a freedom to phase in HD services later without requiring simulcasting, compatibility with CD television is regarded as crucial. Such a system will be introduced in Europe from 1992 and could become established by 1995. Fully digital systems, however, are unlikely to reach a comparable stage of development before the year 2000.

The objective of any broadcast HD service is to provide excellent quality at a reasonable cost to the viewer. It is unclear whether a hybrid or an all-digital approach will eventually have an advantage in terms of overall quality/cost ratio. Early introduction of HD service, however, has obvious advantages and it is thought that there are no major disadvantages with the hybrid approach. All the same, it must be recognized that unless there is a marked difference between HD and CD qualities, growth of HDTV could be inhibited. For an improved television to be future-proof an HD production standard, and to a lesser extent its emission standard need to exceed present-day equipment performance limitations. Considering the additional benefits of compatibility with CD television and hence reduced overall costs, the European system has been chosen to comprise high-definition progressive (HDP) for HDTV production and HDMAC (derived from an interim HDTV standard, HDI) for emission mainly by satellite and cable [2,3]. Description of HDMAC and its transmission performance requirements are the main subjects of this chapter.

3.2. COMPARISONS OF STUDIO COMPOSITE, MAC AND HDMAC BASEBAND SIGNAL FORMATS

The frequency-interleaved composite television systems NTSC, PAL and SECAM generate cross-luminance and cross-colour artefacts which are expensive to remove. These systems are not as readily upgraded, in a compatible manner, to higher definition as those based on time-division-multiplexed analogue components. The MAC/packet (CD) system, first proposed seven years ago for DBS transmission in Europe, has now come to fruition. Signal waveforms of colour-bar pictures for PAL and MAC are compared in Fig. 3.1.

The interval conventionally allocated to horizontal blanking contains a duo-binary data burst of symbol rate 20.25 Mbit/s (equal to the vision signal sampling rate) in the D-MAC variant. It comprises a 6-bit line synchronizing word at the start of each line followed by a data packet for sound and other

COMPARISON OF BASEBAND SIGNAL FORMATS 47

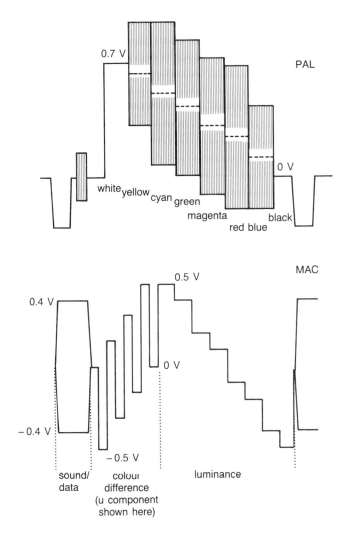

Fig. 3.1 Waveforms for the colour-bar for PAL and MAC.

data services. The total data capacity is nearly 3 Mbit/s, sufficient for eight high-quality sound channels. In another MAC variation, D2-MAC (a subset of D-MAC data multiplex), the symbol rate is halved, hence halving the number of sound channels.

The MAC vision signal is directly related to the CCIR Recommendation 601 4:3 aspect ratio digital-TV studio standard [4]. Colour-difference signals are time-compressed by a factor of 3 and luminance by a factor of 1.5, occupying ⅓ and the remaining ⅔ of the conventional active-line period respectively. The vision signal is preceded by a 15-sample wide clamp region; these together form the analogue signal. The nominal 3 dB bandwidth of the MAC signal is 8.4 MHz, compared with the 5.75 MHz flat bandwidth for luminance in Recommendation 601. Significant differences are that in MAC, the two colour-difference components are transmitted on alternate lines in each field, rather than on all lines. Recommendation 601 signal amplitudes for the 100% colour-bar signal are 700 mV for luminance and ±350 mV for colour difference. MAC amplitudes are 1 V for luminance and the colour-difference components are limited to ±500 mV (from ±650 mV), resulting in a maximum of 98% displayed saturation [5]. The luminance bandwidth after MAC decoding is unchanged from that of composite television, and the available colour-difference bandwidth in both MAC and Recommendation 601 is two to three times greater than PAL. The clamping level for Recommendation 601 is at black level, and for MAC at 0.5 V luminance.

All MAC decoders are pre-programmed to give an alternative set of expansion ratios 33% greater, namely two for luminance and four for colour-difference to allow for 16:9 aspect-ratio reception.

3.3. DESCRIPTION OF HDMAC

The 16:9 aspect ratio HDMAC [6], a new member of the MAC/packet family, is designed to complement existing 4:3 aspect ratio MAC/packet DBS services. MAC and HDMAC multiplexes are compatible with each other and are conveniently related to the 625-line interlaced Recommendation 601 standard in respect of the transmitted scanning raster and sample rates. HDMAC is presently derived from an HDI source (see Chapter 2, Table 2.4) and has twice the luminance sample density of Recommendation 601 in both horizontal and vertical directions, for fully reconstructed static pictures. The colour-difference sample density of HDMAC is half that of luminance in both directions.

3.3.1 Vision and data multiplex of MAC/packet family systems

Signal mapping [7] of the MAC/packet television frame for family members D and D2 is reproduced in Fig. 3.2. For HDMAC, it is supplemented by digitally assisted TV (DATV) data located in the conventional vertical blanking interval. This allows reconstruction to HDI on reception. HDMAC requires a linear-phase 3 dB baseband channel of 11.2 MHz. This compares with both D-MAC and D-2MAC requiring 8.4 MHz to meet the full luminance resolution of Rec. 601 for 4:3 aspect ratio pictures. These MAC systems can also provide a reduced resolution down to that of domestic-grade composite television in the 8.4 MHz bandwidth for 16:9 pictures.

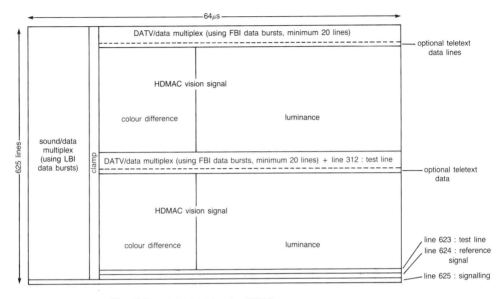

Fig. 3.2   HDMAC/packet TDM structure.

3.3.2   HDMAC vision coding

In order to carry the static resolution of HDI requiring 40.5 MHz (Nyquist bandwidths of 27 MHz for luminance plus 13.5 MHz for colour difference); a reduction by a factor of four is required to transmit this in the 10.125 MHz base bandwidth of MAC/packet system. The transcoding of HDI to HDMAC essentially employs multidimensional filters and some motion tracking. Three parallel coding branches perform different amounts of filtering to allow downsampling by factors of two, four, and eight, respectively in the 80, 40, and 20 ms update modes or branches. Encoder branch selection is made on the basis of least error, in 16×16 pixel blocks over the whole of the locally reconstructed HDI signal. The net result is a constant output rate of 20.25 mega-samples/s. The coding status information is sent over the auxiliary DATV channel, allowing the decoder to act as a slave to the encoder. Additional coding techniques may be employed with the view to maximizing HDMAC quality/decoding-cost ratio, whilst minimizing quality loss of compatible-MAC.

50 TELEVISION OF IMPROVED QUALITY

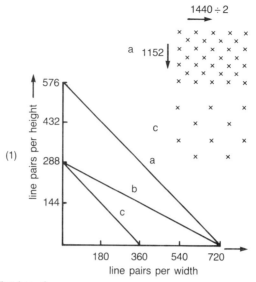

coding branches
a) static   80 ms mode
b) tracked motion   40 ms mode, motion compensated for
                    velocities up to 12 samples per 40 ms
c) untracked motion 20 ms mode, rapid and sudden changes
                    except when in film mode

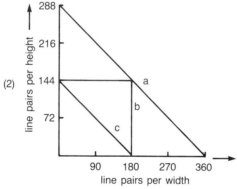

Fig. 3.3 Transmissible range of (1) luminance spatial response and (2) colour-difference spatial response.

In static coding branch (a) of Fig. 3.3, the two-dimensional signal is diagonally filtered, allowing spatial downsampling by a factor of two and providing MAC compatibility. Alternate HDI frames are discarded and the downsampled information sent over two 625-line MAC frames using line-shuffling as illustrated in Fig. 3.4. The complete picture information of the HDI static mode is thus transmitted in 80 ms. Subsampling by two horizontally with an offset of half the sample period in every second and fourth field of the four-field sequence allows transmission of the information over a 10.125 MHz channel.

For pictures containing rapid motion, more severe filtering and sub-sampling is employed, as shown in Fig. 3.3 (c), so that information can be transmitted in 20 ms without the need to discard fields. A consequence of this is that if a highly detailed object starts to move, its resolution is halved. For slowly moving objects, where the eye can track motion, this loss is noticeable. Motion estimation and compensation techniques are therefore employed in the two-field period (40 ms) branch to allow transmission of intermediate levels of detail. For motion tracking, one motion vector is generated for each block and transmitted along with the branch switching information in the DATV data.

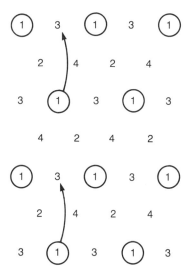

Fig. 3.4 Four-field sequence of static mode and line shuffling.

The use of 625 lines for the transmission of HDMAC allows its reception as compatible MAC on a conventional MAC receiver. In this case the 'line shuffling' causes temporal fluctuations at 12.5 Hz in detailed static areas, a high-frequency dot crawl on horizontal edges, and noise in textured areas. These artefacts are reduced to acceptable levels by a temporal low-pass filter in the static mode. A further vertical low-pass filter is used in the 40 ms mode to minimize the 25 Hz judder on compatible-MAC pictures caused by the conversion. In the HDMAC decoder, two complementary filters are used to restore the signal. The effect of these filters on compatible-MAC is to produce softer but quite acceptable pictures.

3.3.3 Additional processing of MAC/packet signals for emission

HDMAC requires a full theoretical bandwidth, half the MAC vision sampling frequency of 20.25 MHz. This in practice requires Nyquist shaping of the HDMAC spectrum about 10.125 MHz. To minimize the effects of noise and interference in transmission the shaping is shared equally between transmit and receive ends. A $\pm 10\%$ skew-symmetric roll-off resulting in a vestige of 10% has been proposed. In theory this allows tandem connections of an ideal channel of 11.14 MHz, and in practice a linear-phase 3 dB channel of 11.2 MHz. Further processing may be applied to the HDMAC signal to limit the analogue signal amplitude, since a black-to-white input transition of 1 V on encoding produces signal excursions in the region of 1.5 V.

The HDMAC vision signal is subjected to a nonlinear pre-emphasis, E7 [8], before transmission to reduce FM channel noise impairment on reception. In essence the higher-frequency lower-amplitude signals are emphasized before transmission and de-emphasized after reception together with higher-frequency triangular noise of the channel. This technique is especially beneficial for plain areas of the picture where noise is usually most objectionable. Unlike the conventional linear pre-emphasis which is applied to the entire MAC/packet family signal before FM modulation, E7 is applied at baseband to vision signals only. E7 is currently applied to HDMAC only, but will almost certainly be applied to all MAC signals in the future owing to the observed noise improvement of 3 dB for 4:3 aspect ratio and of 1 dB for 16:9, with FM transmission.

The digital techniques inherent in the generation of MAC/packet family signals result in a signal comprising 8-bit words of 20.25 Mega-samples/s (a data rate of 162 Mbit/s), before conversion for analogue transmission. Hitherto, only an analogue interface between MAC encoder and decoder has been possible but, in the future, digital interfaces will also be made available.

3.4. RELATIVE PICTURE QUALITIES

System quality and transmission quality (items (ii) and (iii) in Table 3.1), together with cost, must be taken into account to engineer successfully the co-existence of conventional and improved television. In more detail, these are the relative quality of the two systems excluding transmission impairments, and relative qualities of transmitted and received signals due to transmission impairments alone in the respective systems. These two requirements are independent of each other and are discussed separately in the following subsections.

3.4.1 Quality/cost targets for HD relative to CD television

A distinct increase in subjective quality, combined with a justified increase in cost, is a primary requirement for any improved television system. A 30% increase in quality on the CCIR quality scale (given in Appendix 3.1.3) would be regarded as worth while. Since a viewer experiences quality on a logarithmic scale, it could be said that at least a 3 dB increase in viewing appeal is necessary. This could arguably result from the following objective changes:

- an increase in picture aspect ratio, say from the normal value of 4:3 to 16:9;
- a decrease in viewing ratio (viewing distance in terms of picture height H) from the normal value of 6H to 4H;
- an increase in resolution to maintain the perceived definition.

The enhanced viewing experience is a direct consequence of the increased viewing angles. An increase in viewing angle by a factor of 1.5 vertically is due to the change in viewing ratio, and by a factor of 2 horizontally due to the changes in viewing ratio and picture width. From a system point of view, a 16:9, 625-line, 50 Hz non-interlaced standard with 11 MHz bandwidth (-3 dB) for luminance and half of this value for the colour-difference components would meet the above requirements. Such a solution would require a gross sampling rate of 40 mega-samples/s, and can be referred to as extended-definition (ED)TV.

In contrast, HDMAC, requiring 20 mega-samples/s, is capable of providing a 6 dB increase in viewing experience, assuming availability of suitable displays. This corresponds to a 50% increase in quality on the CCIR quality scale. In this case a further reasonable increase in receiver costs could be justified.

The 30% and 50% quality differences of HDMAC relative to PAL are given in Table 3.1, which also defines the necessary viewing conditions ((a) and (b) relative to (c)).

54 TELEVISION OF IMPROVED QUALITY

Table 3.1
(i) System parameters
(ii) Relative qualities of PAL and HDMAC without transmission impairments
(iii) Relative qualities of received and unimpaired emitted signals in each system

		HDMAC Source target (a)	HDMAC Receive target (b)	PAL Receive target (c)
	Viewing conditions			
(i)	**system parameters**			
	displayed aspect ratio	16:9		4:3
	peak brightness, Cd/m²	>200	70-140	140
	contrast ratio	>50	30	
	viewing ratio, H	3	4	6
	display diagonal, inches	45	32	26
	likely displays	rear projection	direct-view CRT	
	static resolution			
	luminance - vertically	2 times Recommendation 601	Recommendation 601	
	luminance - horizontally	1.5 times Recommendation 601	⅔ of Recommendation 601	
	colour-difference - vert.	½ of the above	½ of the above	
	colour-difference - hor.	½ of the above	¼ of the above	
	display			
	scanning standard	e.g. 1250/100/2:1	1250/50/2:1	625/50/2:1
	resolution (1)			
	required	720	3/4(720)	360
	envisaged	?	480	240
	system artefacts	at threshold of perceptibility value		cross-colour cross-luminance
(ii)	**relative qualities (without impairments)**			
	factor	times 2	times 1.4	times 1
	difference	50%	30%	-
(iii)	**target limits for each impairment (2)**			
	channel noise	threshold value at DBS C/N of 24 dB	grade 4.5 at DBS C/N of 17.5 dB	grade 4 at unified-weighted S/N of 42 dB
	waveform distortion	grade 5 negligible K	grade 4.5 5%K	grade 4 7%K
	rest of impairments	each at threshold	each 3 dB worse than threshold	each generally 6dB worse than threshold
	resulting quality due to noise and waveform distortion	grade 5	grade 4.5 (DBS) grade 4 (DBS + cable)	grade 3.5

Note (1) Line-pairs/picture width, reaching modulation transfer value of −6 dB to −12 dB.
(2) Critical natural pictures

There are two sources of noise present in an FDM, DBS transmission: that due to dish size restrictions, and that due to noise-like inter-channel interference. Both influence the received carrier-to-noise ratio (C/N). The noise limit will typically be 24 dB, perceptibility threshold value for 3H viewing as specified in (a) of Table 3.1. Dish size, however, must be made large enough to make its contribution negligible relative to that due to interference. Clearly, interference noise sets the noise-floor, which would occur with a dish diameter of about 2 m.

The HDP production standard in comparison with HDMAC will rate twice as good (when not limited by cameras and displays which at present fall short of the system potential by a factor of two). The higher quality is not surprising given the luminance bandwidth of 60 MHz for HDP and half this value for colour difference; and the matching display resolution, twice that likely with HDMAC direct-view CRT displays in the domestic environment.

3.4.2 Minimum received qualities of PAL and HDMAC due to transmission impairments

In any broadcast, received quality is a function of cumulative transmission impairments, of which the random noise impairment is usually most dominant. Objectively, the law of addition for noise impairments [9] is such that when a link is divided into two segments, each segment must have a 3 dB better noise performance if the link performance is to be maintained at the original level. If divided into four segments, a 6 dB higher target for each would be required. In the following, the random noise impairment budget is determined for the MAC/packet family reference broadcast chain of Fig. 3.5.

An economically viable approach is to make the video S/N ratios equal for each of the four transmission sections (broadcast network, DBS link, CATV link and receiver/decoder), whilst keeping an overall quality of grade 4.5 (87.5%) on the CCIR five-point quality scale. This occurs at an S/N corresponding to the threshold of perceptability value, that is almost grade 5 for each section. The S/N limit for each section is thus equal to the grade 4.5 figure plus 6 dB.

For DBS the grade 4.5 occurs at a DBS C/N of 19.5 dB for 3H viewing of a particular set of pictures (see section 3.7.1 for derivation). These figures do not include contributions from any other sources. The target grade 4.5 can be realized using 60 cm diameter aerials, for 99% of the worst month at the edge of the service area, where the specified EIRP (effective isotropic radiated power) is -3 dB from the bore-site figure for a full-power DBS.

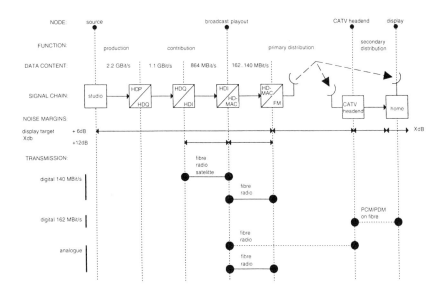

Fig. 3.5 A reference broadcast chain for the MAC/packet system.

For comparison, the normally achieved quality limit for composite television for 6H viewing is nearer grade 4 (75% quality) each for noise and linear waveform distortion, with a resulting total value of about grade 3.5 (62.5 %).

Since noise is the only impairment of significance in DBS, setting the limiting quality target to a one grade higher figure for HDMAC (achievable with a 60 cm dish) represents a significant advance on CD television, especially as it allows a quality grade 4.0 (the current level of CD television) with a relatively inconspicuous 45 cm dish.

3.5. REFERENCE BROADCAST CHAIN AND FURTHER NOISE PERFORMANCE OBJECTIVES

It is usual to define a reference broadcast chain to establish noise performance objectives. A network for MAC/packet signals will contain a mix of analogue and digital links as shown in Fig. 3.5.

Initially the network downstream from the broadcast playout will be as shown. HDMAC links from playout centre to headend of a large cable network could eventually be 140 Mbit/s by fibre and/or microwave, bypassing the DBS link. Digital transmission to the home from a large node in the cable network could in the future employ transcoded PDM [10] for low-cost or PCM for a higher noise performance, both at 162 Mbit/s, allowing respectively analogue or digital interfaces to the receiver.

An HDMAC 140 Mbit/s link between a playout and an uplink represents a further subsection of the chain. Objectively it will need to have 12 dB greater S/N than the required value of 39 dB unweighted for flat noise to ensure quality grade 4.5 into the homes, discounting the improvement from E7. This target has been reached using 7-bit hybrid DPCM/PCM encoding [11].

Fixed HDI links for contribution will be at 140 Mbit/s using DCT (discrete cosine transform) coding, but whether the required S/N (objectively 6 dB above the subjective threshold value for 3H viewing) can be achieved is not yet known. Mobile broadcast links, by comparison, have a practical transmit aerial size limit of 2 – 3 m and will use DCT coding at about 70 Mbit/s. In this case a poorer noise performance down to the threshold of perceptibility value for 3H viewing may have to be accepted, potentially eroding the target quality for such coverage.

3.6. PROCEDURE FOR EVALUATING TARGET CHAIN PERFORMANCE

The following steps should be taken.

- Choice of a subjective method that allows:

    (i) summation of quality due to dominant, unrelated primary impairments and

    (ii) specification of secondary interference impairments, and those resulting from network implementation tolerances.

- Identification of types of unrelated impairments in the complete chain, and determination of their relative influence on performance and whether these fall in categories (i) or (ii) above.

- Recognition of objective parameters directly relevant to quality, and derivation of subjective-to-objective relation for each type of impairment.

- Determination of typical source quality at programme playout centres; this involves a consensus on overall received quality limits to the home for a specified percentage of service-time, with regard to practicalities and costs, and a lower but usable performance limit (say, by one grade) for remainder of the service-time, on account of occasional, noisy programme source or adverse propagation characteristics, with a service failure criterion set at a minimum of half a grade below this value.

- Definition of a reference broadcast chain meeting these quality targets.
- Allocations of an agreed total objective limit for each impairment to individual sections of the reference chain with regard to practicalities and overall costs, using the laws of addition.

The above procedure is applied in the following, for the all-important noise impairment. The last three items above were covered in sections 3.3, 3.4 and 3.5 of this chapter, and some of the remaining impairments of lesser significance were also discussed in an overall sense.

Subjective methods The quality method using non-expert viewers and a continuous grading scale with a five-point quality scale [12] (described in Appendix 3.1.3) is the only method shown to give a reliable summation of performance from dominant, coexisting impairments. It is essential to characterize each of these individually relative to a specified set of impairment-free pictures for the viewing conditions involved. The unimpaired viewing reference here is assumed to have all impairments at, or less than, the threshold of perceptibility values, representing a 100% quality on normalization of mean opinion scores in a subjective test. In the foregoing this method is proposed for the rating of primary impairments, in preference to any other scale, for the reasons given below.

The double-stimulus method with the five-grade impairment scale used within the EBU is not directly suitable for qualitative measure, as it is task-oriented. It generally yields pessimistic results and hence is more suitable for rating of secondary impairments which need to be set to a greater stringency. Unlike the quality scale, the impairment scale has not been shown to give a valid summation of quality from unrelated, coexisting impairments. It is, however, very reliable for determining rank orders of especially small impairments.

The likely transmission impairments, in order of diminishing occurrence and amounts of influence on overall quality in MAC broadcast reference chain, are as follows:

- wideband random noise;
- linear waveform distortion;
- vision-scrambling noise;
- sine-wave interference;
- interchannel and interservice interferences;
- low-frequency interference below 10 kHz;
- line-time nonlinearity;

- differential delay between luminance and colour-difference components;
- digital transmission errors in an increasingly digital environment.

Proposed target limits In a predominantly analogue transmission chain, the primary impairments of noise and waveform distortion are cumulative. It is proposed to set performance limits for these impairments on the basis of quality grade 4.5. The remaining secondary impairments do not accumulate to any great extent but need to be kept relatively small, each ideally set at the threshold of perceptibility value or a little worse as encountered in practice. Setting limits on a threshold of perceptibility basis offers a relatively unambiguous reference for routine use, subsequent to establishing subjective-to-objective relations in formal subjective tests.

The threshold of perceptibility value could be determined as follows. An impairment is repeatedly decreased or increased in small steps by the viewer until the impaired and unimpaired picture qualities are judged identical. For consistent results viewers will need screening for a minimum visual acuity. A total of three or four observers is sufficient to yield reliable results.

3.7. PRIMARY IMPAIRMENTS

3.7.1 Wideband random noise

In a linear baseband system, wideband random noise is conventionally measured in the frequency band from 10 kHz to the upper nominal 3 dB frequency of the video channel. It is expressed as a ratio:

$$\frac{\text{nominal amplitude black-to-white video signal voltage}}{\text{rms voltage of Gaussian noise (S/N) encountered in transmission}}$$

The noise is usually weighted by a filter network that enables subjective quality to be expressed in equivalent objective magnitude, independent of the noise spectrum.

The following gives luminance noise weighting filter time-constants for Recommendation 601, MAC, HDMAC and HDI. Factors taken into account when scaling are the signal compression, picture-aspect and viewing ratios, as proposed for MAC by the UK in the CMTT [13]. The starting point for calculations is the well proven 625-line luminance value of 200 ns originally specified for 6H viewing. The resulting weighting factors for these systems are also given in Table 3.2 for the commonest noise spectra.

60 TELEVISION OF IMPROVED QUALITY

Subjective impairment units [14] (imps) are used as a measure for the primary impairments. In MAC systems the two colour-difference components together introduce as many imps impairment for noise as the luminance component. In objective terms the combined effect is to raise system S/N requirement by 2 dB from that for luminance. This is not unlike the concept of unified noise weighting in composite television, but for component systems it is proposed to base it on the well-proven, traceable and fundamentally correct luminance noise weighting. This approach together with the already established subjective-to-objective relation for 625-line luminance noise [15] allows performance of MAC type systems to be determined. For example, to achieve a quality grade 4.5, S/N weighted equals (41.2 + 2) dB based upon the subjective/objective relationship of Appendix 3.1.2, plus the colour-difference contribution. The corresponding unweighted values (using Table 3.2) for HDMAC viewed at 3H are hence (43.2 − 7.3) dB and 43.2 − 4) dB for triangular and flat noise respectively. The ratios for a grade 4 (75% quality) are 3 dB lower.

Table 3.2 Weighting time-constant (ns)/weighting factor (dB), noise bandwidth (MHz)

	E1 de-emphasized triangular			Flat			3 dB bandwidth (MHz)
	6H	4H	3H	6H	4H	3H	
Recommendation 601				200 6.9			5.6
4:3 MAC	133 12.8	88 9.9		133 6.9	88 5.4		8.4
HDMAC		66 9.2	50 7.3		66 5.0	50 4.0	10.1
HDI					50 6.9	37.5 5.8	22.4
Compatible-MAC	100 10.7	66 7.9		66 4.4			8.4

Where nonlinear signal processing such as E7 is used, there is an apparent subjective improvement equivalent to a 3 dB decrease in S/N for triangular noise and 1 dB for flat noise (after deducting a 2 dB penalty introduced by the MAC compatibility filters in HDMAC). This effect is separate from the weighting mechanism of linear systems. The consequence of E7 is to reduce unweighted S/N for grade 4.5 and 3H viewing to 32.9 dB for triangular, and 38.2 dB for flat noise. A variant of E7 currently under investigation may be used to yield a further noise improvement of up to 2 dB.

The requirements for random noise in DBS to achieve grade 4.5 are summarized in Table 3.3, from calculation using Appendix 3.1.1. In the case of random noise on cable, where the noise spectrum is flat, S/N unweighted of 38.2 dB for quality grade 4.5 HDMAC and 3H viewing equals a VSB-AM C/N of nearly 40 dB per 5 MHz IF bandwidth. The threshold of perceptibility figure will be 6 dB higher.

Table 3.3 Random noise performance of DBS

	E7 pre-emphasis	viewing ratio	S/N unweighted (dB)	C/N (dB)
HDMAC	yes	3H	32.9	19.5
4:3 MAC	no	6H	30.4 (43.2—12.8)	14.7 (30.4—15.7)
16:9 MAC compatible-MAC	yes	6H		~15

3.7.2 Linear waveform distortion of luminance

Since the days of monochrome television, linear waveform distortion has been characterized by a pulse-and-bar waveform rating method termed 'K-rating' [16,17]. This can be directly applied to MAC type systems. In the method, a narrow pulse and a field-bar waveform represent the extremes of picture detail, and a line bar represents an intermediate level of detail. For reasons of test-signal reproducibility and subjectively relevant characterization of channel distortion, the waveform shapes are sine-squared. This shaping can be achieved simply by passing both a rectangular bar and an impulse, of width less than $1/5$ of the required value, through an appropriate passive network. The half-amplitude duration widths T and $2T$ pulses used in the K-rating are given by $T = 1/(2f)$, where f is the ideal channel bandwidth which would approximate to the 3 dB value in practice.

An ideal channel produces insignificant distortion to the $2T$ pulse whereas a T pulse will be distorted by ringing. The filtered T pulse represents the finest detail of subjective importance and the objective limit of the channel. The ringing is of no significance for the recommended viewing ratio as it is imperceptible. Departures in T pulse shape from theory are assessed by $K3$ and $K4$ factors, which in turn relate to subjective quality. $K1$ and $K2$ factors are measured from a $2T$ pulse in composite television, in conjunction with the chrominance/luminance gain and delay inequality measurements. The inequality measurement has hitherto obviated the need for T pulse measurement, but it must be reintroduced for component systems [18].

The weightings inherent in the K-factors serve to equalize subjective effects of distortions (such as echo and blur) to that from an isolated, undistorted echo of $\geq 8T$ duration, as embodied in the $2T$ K-rating graticule. The subjective effects become independent of echo timings of this duration, latterly amended to $16T$. This has enabled $K1$ to be defined as the reference for subjective-to-objective transfer in the K-rating. Its value is expressed as echo-to-signal amplitude in percentage.

A rigorous method of measurement is termed 'acceptance' K-rating. It is accurate but time-consuming. A relatively approximate method termed 'routine' K-rating is therefore normally used, except for critical network designs or commissioning. The normal practice is to take the dominant K-factor from a set comprising a field bar, a line bar and $K1$, $K2$, $K3$, $K4$ as the measure of linear waveform distortion. $K1$ and $K2$ are measured by the conventional method; the routine method for $K3$ and $K4$ measurement is new and is given in Appendix 3.2.

For MAC/packet systems, K-factors of interest are $K3$ and $K4$ for echoes of any duration and bandwidth truncation (blur). Whilst $K1$ and $K2$ are useful for sensing echoes of durations $>2T$ with relative ease, they are insensitive to distortions in the upper half of the channel. The field bar has to be separately K-rated. The line bar does not need to be K-rated as it is supplemented by a full luminance duration bar (detailed later) to meet the stringent requirement of the noise-like impairment caused by interaction of bar-tilt distortion and vision descrambling.

Provided that an automatic clock phase lock is used in HDMAC, the weighting inherent in K-rating for echoes does not seem to differ greatly from that originally derived for monochrome television. The practice of waveform by K-rating method should therefore continue to be used, especially since the same method can be directly used to characterize D-MAC data if echo weightings of $K1$ are ignored. It may be noted that this makes the performance limit for data more stringent than that for D-MAC vision for echoes close to the pulse.

The importance of $K3$ and $K4$ factors in component systems requiring T pulse measurements is not yet widely recognized. They are especially useful for HDMAC vision and D-MAC data characterization, self-evident from the example of linear-phase bandwidth truncations values given for $K3$ in Table 3.4. Similarly, $K4$ can be used to characterize the effects of delay distortion, especially for HDMAC where an interpolation technique is employed.

Subjectively, echoes of 5% K produce quality grade 4.5 for 625-luminance and this seems to apply to MAC/packet signals, including HDMAC. A similar grade on the stringent impairment scale of the EBU would occur at 3-4%K. The same also occurs with a 10-15% linear-phase bandwidth truncation from that of the required 3 dB bandwidth of 11.2 MHz for HDMAC. With HDMAC the limit of perceptibility for isolated, undistorted positive echoes occurs at 16% amplitude for echoes at time interval $2T$, and 8%, 4%, and

2% for echoes at $4T$, $8T$ and $16T$ respectively. These are identical to those for conventional television.

For D-MAC data it is usual to express bit error ratio (BER) against video S/N unweighted in an 8.4 MHz channel. In practice a BER of 10^{-4} is achievable at DBS C/N of 10 dB or S/N unweighted of 26 dB with threshold decoding, after allowing for a 2 dB noise margin on the theoretical figure for data decoding circuits. The available noise margin for this BER at 20 dB C/N is hence 10 dB. This will be totally eroded by a 20% echo, or a 12% echo plus a 15% bandwidth truncation (equal to a 6% $K3$ distortion). Such linear waveform distortion limits, taken together with the stated noise level, ensure a 10^{-4} BER and with it a virtually unimpaired sound.

Since the noise floor in DBS is at about 24 dB C/N and the value taken in the example is 20 dB, additional noise margin of 4 dB is available to absorb unaccounted cable network impairments, such as further noise and interference. At the other end of the scale, an 8 dB C/N on its own will allow just acceptable sound quality at a BER of 10^{-3} with Viterbi decoding [19]. For DATV data without data protection, a 10^{-5} BER is thought to produce little picture impairment, whilst at a lower BER protection may become necessary.

Table 3.4 Routine K-factors for bandwidth truncations of two widely different cutoff rates

Bandwidth truncation (%)	Slow roll-off raised cosine filter P/S = 0.4		Sharp roll-off delay equalized filter P/S = 0.8	
	K1,K2	K3(2)	K2	K3(1)
0	1.4	3.0	0.25	1.0
10	1.6	3.6	0.25	2.0
20	2.0	4.2	0.6	4.0
30	2.6	5.0	1.0	6.0
40	3.3	6.0	2.0	8.0
50	5.0	8.0	4.5	11

Note 1 Agrees closely with amended acceptance $K3$ (not given here)
 2 Agrees closely with amended acceptance $K3$ up to 20% bandwidth truncations, beyond which the routine figure is less than the relatively accurate acceptance figure, apart from the scaling factor amendment (given below).

Note that the 3 dB linear-phase signal bandwidths (f) for MAC and HDMAC are respectively 1.5 and 2 times 5.6 MHz. Hence, for HDMAC the half-amplitude width of T pulse is 45 ns ($1/[2f]$). P/S is the ratio of 3 – 6 dB bandwidth to 40 dB bandwidth.

For echoes, mid-opinion subjective rating occurs at $K1, K2 = 11K$ [20]. The same has also been reported for bandwidth truncation [18]. Objectively, $11K$ also occurs at a 50% bandwidth truncation of type 2, by the routine method of measurement proposed for $K3$, whereas $K3$ by the acceptance (i.e. the time series) method gives $16K$ for the 50% bandwidth truncation. In order to rectify the discrepancy between the two methods of measurement, it is proposed here to scale the figures obtained by the acceptance method by a factor 11/16.

3.8. SECONDARY IMPAIRMENTS

3.8.1 Vision scrambling noise

This is a new impairment caused by the interaction between line-time waveform distortion (line tilt) and descrambling of 'single and/or double-cut and rotate' vision scrambled signals. Noise will also be generated by interaction between jitter on the recovered clock and the descrambling. The impairment in both cases is not unlike flat random noise except that in the former case it is concentrated at the lower end of the frequency spectrum and is therefore more objectionable.

The impairment imposes limits on permissible line tilt, and clock jitter which is yet to be assessed. The latter can be characterized by monitoring distortion of an active-luminance bar, preceded by black level. The line-tilt is then expressed as a percentage of the bar amplitude. It is normally necessary to omit about 0.5 μs periods at the beginning and end of the bar, so as not to be misled by overshoots on the waveform.

There is no mechanism in FM transmission that can produce line tilt except in baseband circuits between scrambler and modulator and between demodulator and descrambler. If these signals are later subjected to baseband or AM transmissions, additional waveform distortions are inevitable.

The 'double-cut' scramble noise impairment from a 3% linear line tilt in HDMAC subjectively equates to a random noise impairment 3 dB greater than the threshold of perceptibility value of 46 dB S/N unweighted, for flat noise, colour-bar picture and 3H viewing. Alternatively, the impairment is 3 dB less than that for quality grade 4.5 for noise on the same picture. For exponential tilts larger amplitudes can be tolerated for the same amount of subjective impairment.

An objective 3 dB margin about this level of impairment is desirable, thus allowing as much impairment again from interaction between a clock jitter and descrambling. The resulting overall limits for tilt amplitudes are 2, 3 and 4% respectively for time constants greater than or equal to the luminance bar duration, greater than or equal to half the bar duration, and less than half the bar duration.

3.8.2 Baseband protection ratios for sinewave interference

This is usually expressed as $20 \log_{10}$ (p-p vision signal/p-p sinewave interference). The perceptibility threshold values for HDMAC at 3H or 4H (within experimental tolerance) are 63 dB in the range up to 2.5 MHz, above which they decrease linearly to 45 dB at 5.5 MHz and from there down to 30 dB at 12 MHz. A suitable overall limit for a sinewave interference could be 3 dB worse than these values.

It is worth noting that sinewave interference producing only the vertical patterning is converted into random noise-like interference. This would permit a 10 dB greater interference level but at the expense of a random-noise penalty.

3.9. THE POSSIBLE FUTURE OF IMPROVED TELEVISION

Introduction of HDMAC service is expected by 1995 in Europe, but before that date a 16:9 MAC is likely to be introduced, allowing cropped or letterbox format pictures on 4:3 CD displays. New 16:9 MAC receivers will follow, allowing downward-compatible reception of HDMAC transmissions. Small size 16:9 MAC receivers up to a 26 in diagonal, used for viewing ratios ≥ 6H, will not require HDMAC decoding, and could continue to exist indefinitely. They are of vital commercial value, especially since the number of receivers per household is increasing and a large proportion of these are of small size and could continue to be small. Only the main receiver need be of a large size, and include a HDMAC decoder. Compatibility allows HDMAC to be introduced at a pace that could be afforded, whilst minimizing the risk to service providers. Without such assurances, investment into consumer products and the necessary infrastructure for an HD television service would not occur at a pace necessary for a healthy and rapid growth. The European approach to HDTV is well placed because of the availability of DBS media and the foresight in the development of the fully engineered MAC/packet system.

Another prospect for an all-digital HDTV service is now on the horizon. This approach, however, requires simulcasting in CD to serve existing viewers. Digital techniques using DCT are now well advanced and are capable of yielding impressive results, but it is not yet known what quality can be achieved over the 'taboo' terrestrial channels proposed for the USA, at what cost and when. It has been shown that a doubling of quality between HD and 4:3 CD television approaches the discerning viewer's needs. Such an increase is necessary for an HD emission to be future-proof, especially since the HDP production standard is capable of similar additional increase in quality. The latter is, however, suited to community viewing requiring larger-screen presentations. An intermediate quality level for another coexisting HD broadcast service would be worth while and could become available if, as anticipated, wider-bandwidth DBS channels are allocated in the 22 GHz satellite band by WARC 1992. Such a service will almost certainly be all-digital at 30 to 50 Mbit/s [21] and could become available at affordable prices from the year 2000.

Yet another strong possibility for introducing an all-digital service is to supplement the 4:3 CD television with an improved 16:9 ED version in the UHF band using vacant channels, thereby providing extra channels and the possibility of phasing out analogue channels in the more distant future. This approach offers an excellent opportunity to unify the converging requirements of both telecommunications and broadcasting [22,23] industries towards a 10–20 Mbit/s data rate for CD to ED television. These systems are currently being studied by both industries.

In view of the anticipated trend towards versatile displays, serious consideration could also be given to a coexisting EDTV based on a sequential scanning, at least for the source and display, at 20 Mbit/s, and HDTV (HDI) at twice this rate. These standards respectively could be broadcast in existing narrowband satellite channels at 12 GHz and wideband satellite channels at 22 GHz.

APPENDIX 3.1

3.1.1 Equations for FM C/N to video S/N calculation

Luminance channel signal/unweighted noise (S/N) ratio and weighting factors (i.e. the difference between weighted and unweighted values) in an FM system can be calculated for the viewing distance, noise spectra, aspect ratio and signal expansion involved as follows:

Demodulated luminance S/N,

$$N = C/N + 20 \log (Km . L . \sqrt{(B)}) - 10 \log Int \text{ (dB)}$$

where:
    Km discrimination sensitivity (13.5 MHz/V)
    B receiver IF bandwidth (27 MHz)
    L demodulator signal amplitude (1V p-p)
    C/N ratio of rms-carrier to rms-noise in dB measured in bandwidth B

APPENDIX 3.1

Int is the total noise power in base bandwidth (f_2) of the demodulated FM signal modified by noise-weighting and MAC de-emphasis, E1.

For a simple demodulator,

$$Int = \int f^2 \cdot |S(f)|^2 \cdot |T(f)|^2 \, df$$

where:
 f is frequency
 $T(f)$ is the transfer function of the noise-weighting network
 $S(f)$ is the transfer function of the de-emphasis network

and are given by:
$$|T(f)|^2 = 1/1 + (2 \cdot \pi \cdot v \cdot f)^2$$
$$|S(f)|^2 = G(1 + X^2 \cdot B \cdot f^2)/(1 + X^2 \cdot C \cdot f^2)$$

where:
 X is signal expansion ratio
 f_2 is 3 dB noise bandwidth
 $f_1 = 0.01$ MHz
 v is noise-weighting time-constant where $v=0$ for no weighting

For MAC de-emphasis,

 $G = 2$
 $B = (1/1.5)^2$
 $C = (1/0.84)^2$

The luminance S/N equation is valid above the so-called noise threshold, which occurs below 12 dB C/N with conventional discriminators. At these C/N ratios additional noise mechanisms occur producing additional subjective effects.

3.1.2 Subjective rating of 625-line luminance noise

The recommended relationship between subjective impairment (I_L) in imps and weighted S/N (N) in dB for 625-line luminance signals for 6H viewing is as follows. It can be directly applied to MAC systems.

$$I_L = (d/0.0184)^{2.61} \text{ (imps)}$$

where:
 d = weighted rms noise voltage/peak-peak picture voltage = antilog $(N/20)$

3.1.3 Relationship between quality scale, normalized opinion score and subjective impairment units (imps)

Normalized mean score (U)	CCIR 5 point quality (Q)		Impairment (imps) (I)
1	5	excellent	0
0.75	4	good	0.33
0.5	3	fair	1
0.25	2	poor	3
0	1	bad	infinite

Note $I = (1/u) - 1$
$U = 1/(I+1) = (Q-1)/4$
$Q = (I+5)/(I+1)$
$I = (5-Q)/(Q-1)$

3.1.4 Summation of subjective impairment units

Provided that the proposed single-stimulus subjective method and quality scale is used for the assessment of each of the impairments in turn, the following empirical relationship has been shown to apply:

$$I_{total} = I_n$$

where:

$I_n = [(100/Q^n) - 1]$
I_n is impairment unit in imps for each impairment
Q_n is quality in percentage relative to 100% quality at threshold of impairment perceptibility

Total quality is then given by the inverse transform:

$$Q_{total} = [100/(I_{total} + 1)] \ (\%)$$

APPENDIX 3.2

3.2.1 Measurement of $K3$ (T pulse-to-bar ratio)

In order to measure $K3$ and $K4$ it is necessary to employ a high-quality phase-corrected low-pass measurement filter to simulate an ideal transmission channel. This is conveniently provided by employing two filters in tandem, each of which conforms to the specification in CCIR Recommendation 601

[24] for a luminance channel filter, but with the frequency scaled by a factor of 11/6 for HDMAC. The T pulse-and-bar waveform is passed through the measurement filter and the T pulse-to-bar ratio at the output of the filter is adjusted to unity by adjusting the generator pulse height. This waveform is applied to the system under test and the pulse-to-bar ratio measured. $K3$ is given by ¼ of the percentage change in the ratios between the two conditions — filter alone and filter plus system under test.

3.2.2 Measurement of $K4$ (T pulse shape)

The T pulse-and-bar waveform is passed through the measurement filter and the first pre- and post-overshoots are measured as a percentage of the pulse amplitude. The system under test is then introduced in tandem with the measurement filter and the overshoots again measured as a percentage of the pulse amplitude. $K4$ is given approximately by ⅓ of the percentage change in pre- or post-overshoot (whichever is the larger).

ACKNOWLEDGEMENT

The author gratefully acknowledges the assistance given by Peter Elmer of Visual Telecommunications in preparing the final version of this chapter.

REFERENCES

1. Ninomya Y et al: 'A Single Channel HDTV Broadcast System — The MUSE', No 304, (Sep 1984).
2. Searby S: 'Broadcast TV systems', in Kenyon N D ed, 'Audiovisual Telecommunications', Chapter 2, Chapman & Hall (1992).
3. Haghiri M R and Vreeswijk F W P: 'HDMAC coding for MAC compatible broadcasting of HDTV signals', IEEE Transactions on Broadcasting, 36 , No. 4 (Dec 1990).
4. CCIR, 'Encoding parameters of digital television for studios', Recommendation 601-1, Geneva, (1986).
5. Windram M and Morcom R: 'MAC colour-difference signals', IBA Report 124/83.
6. System 2 in the CCIR Report 801-3, Part 7.
7. CCIR Report 1073.

8. Beech B and Baudouin C: 'Compatible non-linear pre-emphasis for MAC signals', IBA Report 141 (1989).

9. CCIR Recommendation 567.

10. Burgess G D: 'A pulse density modulation video codec', BT Technical J, 9, No 2 (Apr 1991).

11. Input document to CMTT/EWP-2, CCIR Study Groups (1986-1990).

12. CCIR Recommendation 500.

13. CMTT/1047-E.

14. Allnatt J W: 'Subjective rating and apparent magnitude', Int J Man Mach Stud, 7 (1975).

15. White T A and Reid G M: 'Quality of PAL colour television pictures impaired by random noise: stability of subjective assessment', Proc IEE, 125, No 6, pp 571—580 (1978).

16. Lewis N W: 'Waveform responses of television links', Proc IEE, 101, Part 3, No 72 (1954).

17. Macdiarmid I F: 'Waveform distortions in television links', Post Office Electrical Engrs J, 532, Pts 2 and 3 (1959).

18. CCIR Recommendation 567-2, Part C, Annex IV.

19. Viterbi A J: 'Convolutional codes and their performance in communication systems', IEEE Trans Commun Technol, COM-19(5,2), pp 751—772 (1971).

20. Allnatt J W: 'Subjective quality of television pictures impaired by linear waveform distortions', Electronics Letters, 13 (1977).

21. Cominetti M: 'Perspectives of digital television HDTV by satellite', IEE Colloquim on Prospects for digital television broadcasting, Digest 1991/037.

22. Parke I and Morris O J: 'International standards for digital TV coding', IEE Colloquim on Prospects for digital television broadcasting, Digest 1991/037.

23. Mason A G and Lodge N K: 'Digital television transmission in the terrestrial UHF band', IEE Colloquim on Prospects for digital television broadcasting, Digest 1991/037.

23. CCIR Recommendation 601, Annex 3, Fig 1.

4

A NICAM DIGITAL STEREOPHONIC ENCODER

G T Hathaway

4.1. INTRODUCTION

Broadcast television has been regularly improved over the years. 405-line transmission gave way to 625 lines; monochrome transmission was replaced by colour and teletext services were added, but very little was done to improve the sound. Undoubtedly the quality of the sound transmission has steadily improved and for most viewers the main limitation in the quality of the sound reproduction is in the television receiver. Many have used hi-fi systems to reproduce the TV sound channel and can attest to the good quality of the transmitted sound. However the advent of the compact disc with its high sound quality and a general improvement in the standard of normal domestic hi-fi systems has led to a market for better audio performance from television receivers and also for stereo sound.

In Germany, similar requirements for stereo sound were met by the use of a second analogue sound channel, but tests by the BBC and IBA showed that a digital transmission system was capable of better performance and resulted in reduced interference both to the sound and to the vision. The result was the NICAM 728 standard, NICAM being an acronym for 'near-instantaneously companded audio multiplex'.

Similarly a requirement for broadcast-quality links from the studio to FM stereo radio transmitters led the BBC to develop another digital audio transmission system called NICAM 3. This standard, which is described below, was specifically designed to allow the transmission of three stereo pairs over a 2.048 Mbits/s data link or an analogue video link. Similar requirements

72 A NICAM DIGITAL STEREOPHONIC ENCODER

exist for audio transmission within cable TV systems and there is a wide variety of other applications where a number of audio channels must be transmitted economically over digital or video networks.

4.2. NICAM 728 AND NICAM 3 CODING FORMATS

4.2.1 NICAM 728 format

This broadcast TV system [1] uses a 32 kHz sample rate with 14-bit initial quantization and compression to ten bits on a 1 ms block basis hence the term 'near-instantaneous companding'. A single parity bit is added to each sample and the compression range signalled by modification of the parity bits. Each of the three bits which represent the range code is added, modulo 2, to nine of the parity bits so that 27 of the 32 parity bits in each block for both left and right channels are modified. The decoder uses majority logic to decode correctly both range and parity bits. Bit interleaving is used for burst error protection and the resulting bit stream is scrambled with a pseudorandom binary sequence for energy dispersal. Transmission is by differential quadrature phase shift keying of a carrier placed 6.552 MHz above the vision carrier at a level 20 dB lower. Whilst the normal application of this standard is to carry either a stereo pair or two mono channels for dual language use, etc, it should be noted that one or both of the two mono channels can be replaced by a data channel.

The derivation of the 728 kbits/s data rate, from which the name of this standard is taken, can be understood by considering the data carried in the stereo audio mode in a 1 ms block. The block begins with an 8-bit frame alignment word, followed by five control bits used to signal the audio and data modes and 11 additional data bits used for transmitter control, etc; the audio information totals 704 bits, consisting of 32 samples from the left channel and 32 samples from the right channel, each compressed sample having ten data bits plus one parity bit.

Details of the frame format for the stereo audio mode are shown in Fig. 4.1.

NICAM 728 AND NICAM 3 CODING FORMATS 73

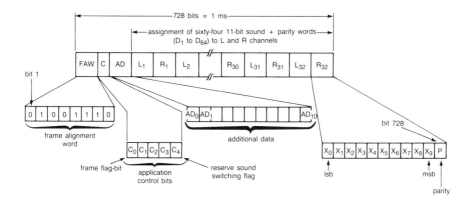

Fig. 4.1 The frame format for the stereo audio mode (reproduced by kind permission of the BBC).

4.2.2 NICAM 3 format

To enable the data from three stereo channels to be multiplexed into one 2.048 Mbits/s channel requires a channel bit rate below 728 kbits/s. This can only be achieved by a reduction in the level of error protection, but as this format is designed for transmission over data links, which have a better error performance than carrier transmission, this is quite acceptable. For the same reason no scrambling is performed and the bit interleaving arrangement is much simpler. NICAM 3 [2] uses the same 32 kHz sample rate, 1 ms block size, 14-bit initial quantization and compression to ten bits but adds only one parity bit for three samples rather than one for each sample. In order to reduce the overheads further the NICAM 3 coding scheme transmits a combined range word for each 3 ms. This unfortunately increases the encode-decode delay and also increases the sample storage requirements but it does allow the words protected by each parity bit to be spread further apart to give increased burst error protection.

The result of the above is that calculation of the data rate must be made over a 3 ms period but as left and right channels are handled quite independently before the final multiplexing it is only necessary to consider one audio channel.

74 A NICAM DIGITAL STEREOPHONIC ENCODER

The important audio data thus consists of 96 samples where each compressed sample has ten data bits, with parity bits adding a further 32 bits. This gives 992 bits to which must be added four additional data bits, used for transmitter control, etc, 11 bits used to signal the Hamming protected compression range code and a 7-bit frame alignment word. The total is thus 1014 bits for 3 ms which equals 338 bits/ms for one channel or 676 for a stereo pair.

Three such pairs give a multiplex data rate of 2028 bits/ms to which are added a further signal bit, 12 justification bits and a 7-bit multiplex frame alignment word. The resulting 2048 bits/ms matches the 2.048 Mbits/s standard primary access channel exactly. HDB3 coding is used to give a standard G.703 interface but at a reduced level the same signal can alternatively be carried over standard video channels.

Figures 4.2 and 4.3 give details of the individual channel format and the six-channel multiplex format for the stereo audio mode.

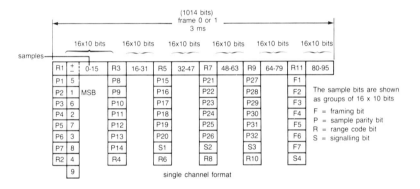

Fig. 4.2 Individual channel format for the stereo audio mode (reproduced by kind permission of the BBC).

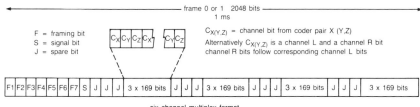

Fig. 4.3 Six-channel multiplex format for the stereo audio mode (reproduced by kind permission of the BBC).

4.3. REQUIREMENTS

Since NICAM 728 encoding is normally used only for TV broadcast transmissions the market for encoders will be limited. Designs developed by the broadcasters are available from test equipment manufacturers but they are expensive because of the small production quantities.

Until recently the need for encoders in cable TV systems has not been fully recognized. At a cable TV headend off-air broadcasts can be received complete with the NICAM carrier, but satellite transmissions use an analogue audio transmission system and in a similar way to local programmes must have their sound encoded in NICAM format if they are to be received in stereo by NICAM-equipped receivers. Since each headend may need 20 or more encoders the requirement for low cost is of paramount importance.

Despite the low-cost criterion, any system to be considered must be capable of meeting the full broadcast specification [1]. This has been achieved by the designs to be described; the only compromise which has been made is in the area of the quadrature phase shift keying (QPSK) modulator and sideband shaping where the requirements for a cable TV system are less stringent than for broadcasting.

NICAM 728 decoding may also be required in a cable TV system in order to be able to separate audio and video, for example to enable channels to be passed through a mixing desk to allow the addition of local advertisements or messages. For locally generated programmes, sound and vision could be kept separate, but for remotely provided channels it is convenient to be able to handle the sound and vision together as far as the mixing desk. Current monaural systems must therefore eventually be replaced by stereo equipment which includes decoders as well as encoders. The interest in NICAM 3 transmission equipment for systems such as the Westminster Cable TV network is centred on its ability to replace existing more expensive and less efficient arrangements currently in use for transmission of stereo audio between the various parts of the system. To suit this application both encoders and decoders are required and performance must again be to broadcast standard.

4.4. CHOICE OF TECHNOLOGY

4.4.1 NICAM 728 and NICAM 3 encoding

In preference to designing application-specific integrated circuits, a digital signal processor (DSP) is chosen, since it provides a more flexible solution as well as achieving the required packing density. DSPs are basically high-speed microprocessors with an instruction set optimized for the implementation of digital filters, etc. In this application, the processing requires a cycle time of under 100 ns, only recently achieved in mass-produced devices. For compactness it is also important that the device has sufficient internal random access memory (RAM), one serial input port and two independent serial output ports.

For NICAM 728 sufficient RAM is needed to hold at least three blocks of 32 left- and 32 right-channel samples for implementation of the dual monaural audio transmission mode which requires two successive 1 ms blocks to be processed whilst a third is being stored. In practice four blocks are more convenient with a consequent input buffer requirement of 256 words. Only two output blocks are required but each must hold 91×8-bit words to give the 728 kbits/s data rate. Rounding these output block sizes up to 128 words gives an output buffer requirement of 256 words, as for the input buffer. The total RAM requirement is thus around 512 words since spare space in the output buffers can be used for temporary variables, etc. This size of RAM is available with several currently available devices but an option to use a further 1 K words allows differential encoding to be performed in software by means of a look-up table with a consequent reduction in peripheral circuitry and improved flexibility in the design. It is true that the look-up table could be stored in EPROM but storage in RAM gives essential speed advantages since the DSP can read program instructions from EPROM and data from RAM simultaneously. The NICAM 3 RAM requirement is similar, though the availability of up to 1 K words allows for a more logical buffering arrangement.

The serial input port requirement is equally important because the analogue to digital (A/D) converter chosen has no parallel interface and board space is at a premium. Similarly the output from the encoder prior to modulation is a serial bit stream, which suggests a requirement for at least one serial output port and also for a port which can operate in a continuous mode. Availability of two independent serial output ports allows differential encoding to be performed in software to simplify the peripheral circuitry and also means that this facility can be easily disabled to provide straight data for applications not requiring differential modulation.

The final choice of DSP is the Texas Instruments TMS320C30 which offers all the facilities required, but other aspects of the hardware design are also important if the goal of fitting the encoder on a standard length single height Eurocard is to be realized. In particular the choice of A/D converter is critical since normal analogue input anti-alias filtering is known to be relatively bulky. Use of a delta-sigma A/D converter simplifies the audio input filter requirements and makes it possible to achieve the required performance in the space available. The converter chosen samples the audio input at 64 times the required rate of 32 kHz but uses only 1-bit quantization. The sampling and quantization are followed by a three-stage finite impulse response low-pass digital filter and decimation to give a full 16-bit accuracy. The result of the oversampling is that the first alias does not occur until almost 2 MHz, where there is negligible input energy, so steep analogue anti-alias filters are no longer required. Overall audio performance is impeccable with 94 dB signal-to-noise-ratio and 0.0015% total harmonic distortion. The filter also has linear phase, a 0.001 dB passband ripple and 86 dB stopband rejection.

4.4.2 NICAM 3 decoding

NICAM 3 decoding is not available in a highly integrated form and again the DSP is used. The hardware design is very similar to that of the encoders. The use of four times oversampling, with digital filter and D/A devices designed for compact disc decoding, simplifies analogue filtering and gives the best possible performance within the same size constraints.

4.4.3 NICAM 728 decoding

By contrast, NICAM 728 decoding is becoming a standard requirement of modern television receivers and so, naturally, highly integrated systems are available to perform this task. Digital-to-analogue circuitry is very similar to that used in the NICAM 3 decoder but in order to maintain compatibility with the existing monaural system it is also necessary for the decoder to include an FM demodulator and automatic audio switching between the two systems.

4.5. HARDWARE REALIZATION

Figures 4.4 – 4.6 show the schematic forms of an encoder for both formats, and decoders for NICAM 728 and NICAM 3 respectively. Much of the operation will be evident from these diagrams, but some specific aspects of interest are discussed here in more detail.

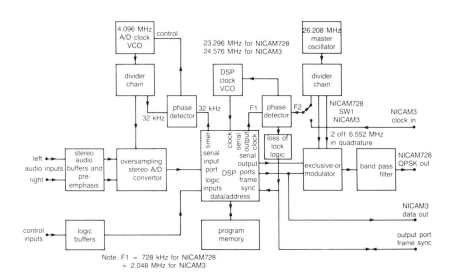

Fig. 4.4 NICAM encoder with particular emphasis on the clock generation, and the phase-lock loops.

4.5.1 NICAM 728 encoder clock synchronisation

One of the main complications with the encoding of NICAM 728 is the awkward relationship between the various frequencies involved — the 728 kHz data rate, the 32 kHz sampling rate, the A/D converter 4.096 MHz master clock and 2.048 MHz serial data clock and the 6.552 MHz carrier frequency. Clearly the 728 kHz data rate and 32 kHz sampling rate must be locked together but a further requirement is that the A/D converter master

HARDWARE REALIZATION 79

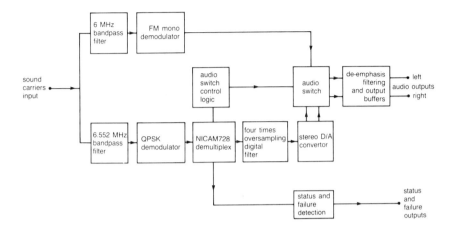

Fig. 4.5 NICAM 728 decoder.

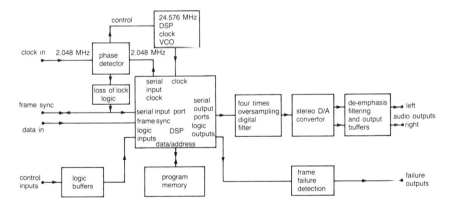

Fig. 4.6 NICAM 3 decoder with special emphasis again on the clock generation, and the phase-lock loops.

clock and sample rate clock must be derived from a single source to maintain a very high timing accuracy.

The DSP serial ports can all run with internal or external clocks so a variety of approaches is possible. A master clock with dividers to give all the above frequencies would have to be at the unreasonably high rate of 3.354 624 GHz — therefore at least two clocks must be used and phase-locked together at a frequency which is a common factor to both. In order to minimize hardware divider circuitry it is also appropriate to use the DSP, as far as possible, for frequency division. A DSP clock frequency of 32 times 728 kHz is therefore chosen. This allows simple derivation of the 728 kHz output data clock and of a 32 kHz waveform from one of the DSP timers. Phase locking of this to the same frequency derived from a 4.096 MHz A/D master clock allows the DSP and A/D converter clocks to be locked together at the sample rate. These frequencies are then all locked to the QPSK carrier by operating the DSP clock in a phase-lock loop which compares a 728 kHz waveform derived by division by nine from the 6.552 MHz carrier with the 728 kHz output data clock derived within the DSP (Fig. 4.4).

For NICAM 3, the frequency relationships are all simple powers of two. The A/D converter clocks could thus have been derived directly from the DSP, but in order to allow the use of the same hardware, the A/D clocks are still produced independently and phase-locked to the DSP. Phase-locking of the DSP to a master clock is still required in order to allow the whole encoder to be locked to an external clock which is essential for NICAM 3 operation where three encoders must be synchronized together. In this case, however, the master clock is fed into the encoder card from an external source.

4.5.2 NICAM 3 multiple cards synchronisation

For correct NICAM 3 operation, bit synchronization of the serial output ports of the three encoders is vital. This is achieved by connecting together the serial port frame synchronization lines for the three cards and allowing each card to release the line when ready. The last one to release the line starts the data stream synchronously for all three cards. The outputs of the cards are then added together using the logic function 'and' before encoding in HDB3 (high density bipolar) format. In order to avoid the disadvantages of assigning one card to be master, the HDB3 encoder also provides a 2.048 MHz master clock to which all three encoders lock as described above. In a similar way the decoder input ports are started synchronously and the DSP clocks locked to the recovered data clock from the HDB3 card.

4.5.3 Differential quadrature phase shift modulation

Modulation of the NICAM 728 bit stream on to the 6.552 MHz carrier is greatly simplified by the software differential encoding as all that is needed then is for two balanced modulators working with carriers in quadrature. The latter are easily produced by digital division of a master oscillator running at four times the carrier frequency and balanced modulation can be achieved digitally by an exclusive-or gate. However, the question of spectrum shaping must be considered. The 100% cosine roll-off specified by the NICAM 728 standard (shared equally between transmitter and receiver) would require special low-pass filtering of the differentially encoded waveforms and analogue balanced modulation. However the exclusive-or system used, as outlined above, does produce a very similar spectrum in the form of a $\sin(x)/x$ function. Sidebands beyond the first nulls in the spectrum must be removed but the result (see Fig. 4.7) is then satisfactory for all but professional broadcast operation. The filter which removes the unwanted sidebands very conveniently has traps at either side, one of which falls at about 6 MHz with the benefit of reducing the level of interfering sidebands around the critical region of the normal FM mono sound carrier. The resulting eye-height is slightly lower than the optimum achievable with true 100% cosine roll-off but due to amplitude errors rather than phase errors. Decoder clock timing tolerance is thus not degraded.

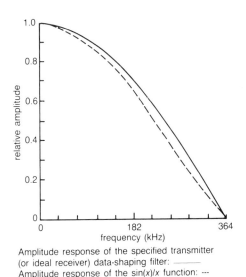

Amplitude response of the specified transmitter (or ideal receiver) data-shaping filter: ———
Amplitude response of the $\sin(x)/x$ function: ---

Fig. 4.7 Amplitude responses.

Figure 4.7 shows the small difference in amplitude response between the $\sin(x)/x$ roll-off and the transmitter contribution to the 100% cosine roll-off. The phase responses are identical.

4.6. SOFTWARE IMPLEMENTATION

Input and output from both NICAM encoders and from the NICAM 3 decoder are in serial form and handled by interrupt routines. In the case of the encoders, the routines are triggered by the DSP serial output ports, and in the case of the decoder by the input ports. Because of the fixed timing relationship between input and output both can be handled by the same interrupt routine.

Investigation of the processing requirements for most of the software routines indicated that there would be little difficulty in keeping within the time allocated, but care was needed when designing and coding those routines which operate at the bit level such as bit interleaving. Assembly of the data into and disassembly from the channel and multiplex formats for NICAM 3 was also found to be more difficult than expected because the bit groupings were not simple multiples of the DSP word length. The final design solves the problem by performing as much processing as possible at the word level with the minimum of bit handling necessary to produce the required result.

4.6.1 NICAM 728 encoder

The NICAM 728 data is put in the format described above, and each word of the output data translated to two differentially encoded words by means of a look-up table before transmission at 364 kbits/s, i.e. at half the total data rate, to the two modulators.

Although mono and stereo channel formats are slightly different, in both cases millisecond blocks are handled independently, so the main body of the software repeats once per block and performs the following functions after initialization.

- Find maxima for both left and right channels.
- Assign scalefactors.
- Compand each sample from 14 bits to 10 bits.
- Calculate even parity on top six bits only of each sample.
- Signal scalefactor in parity, by modulo 2 addition of scalefactor bits to selected sample parity bits.
- Interleave data so that bits from any one audio sample placed at least 16 bits apart in final bit stream.

- Scramble data by modulo 2 addition of pseudorandom binary sequence, for energy dispersal.
- Add additional data bits.
- Add frame alignment word.
- Differentially encode data ready for output.

4.6.2 NICAM 3 encoder

The NICAM 3 channel structure is entirely different from that used for NICAM 728, as evident from Figs 4.1 and 4.2; so, although the basic processing is similar, most of the routines are new. Additional processing is also apparently required to transfer the data to the final multiplex format. In fact, the audio data could have been put directly into the multiplex format, but use of the single-channel format as an intermediate stage was chosen so that the routines developed could also be used for decoding simply by reversal of their function and order.

Since parity and range code transmission operate over a group of three consecutive millisecond blocks, which consitute the stereo pair channel frame, some operations can only be performed once every 3 ms. However, in order to make best use of the available processing power, it is necessary to ensure that as many operations as possible are carried out on a per-millisecond basis. The NICAM 3 encoder software thus performs the following operations every millisecond after initialization except where otherwise stated.

- Find maxima and assign range codes.
- Compand each sample from 14 bits to 10 bits.
- Calculate odd parity on top five bits only of each sample.
- Interleave bits of each sample in the order 9/4/8/3/7/2/6/1/5/0 for burst error protection.
- Assemble and insert housekeeping words (every third millisecond only):
  — combine and interleave parity bits;
  — combine and Hamming encode range codes for three blocks;
  — add four data bits per channel;
  — add channel frame alignment word.
- Multiplex left and right channels using only every third bit position in the output stream.

4.6.3 NICAM 3 decoder

The NICAM 3 specification does not guarantee the fixed relationship achieved in the encoder described here between the start of the multiplex frame and the individual channel frames. Thus for full compatibility, the decoder must be able to achieve multiplex and channel frame alignment independently. The intermediate channel format, used by choice in the encoder, clearly cannot be avoided for the decoder, but its use in the encoder, as explained above, has significantly reduced the total development time. Apart from the need for frame alignment some other software differences are inevitable: for example, the decoder does not need to find the maxima in each block, but it does need to compare the transmitted parity bits with those calculated from the received samples in order to be able to indicate errored samples for correction by interpolation in hardware.

4.7. CONCLUSION

The outline hardware and software design of a pair of digital stereo audio encoders and decoders have been described. Particular mention has been made of those features which have led to the high level of flexibility which has been achieved.

ACKNOWLEDGEMENTS

The author would like to thank the British Broadcasting Corporation for permission to reproduce Figs 4.1, 4.2 and 4.3.

REFERENCES

1. 'NICAM 728: Specification for two additional digital sound channels with System I television,' BREMA/IBA/BBC, approved by the DTI (August 1988).
2. BBC Designs Department Handbook No 6.239(83): 'NICAM 3 Specification' (1983).

5

VIDEOPHONY

I Corbett

5.1. INTRODUCTION

Videophony is a companion audiovisual service to videoconferencing. It may be distinguished by the following features.

- It is primarily for person-to-person rather than group-to-group audiovisual communication.
- It is usually an on-demand service provided on customer-switched networks, whereas videoconferences (like face-to-face meetings) can be scheduled in advance and may use fixed links.

Like all telecommunications services, the need for international standards is paramount to enable the service to be offered on a worldwide basis with guaranteed interworking between terminals supplied by different manufacturers. And it is not only the video coding and audio coding algorithms that need to be standardized, but the system aspects including 'handshaking' protocols when a videophone connection is established, the framing of the data for transmission and the interface to the network. A detailed description of the standards for audiovisual services is given in Chapter 7. This chapter concentrates on the design considerations for videophone terminals, taking into account the requirements for videophone service on a number of customer-switched telecommunications networks. The network most likely to be used for such a service is the new integrated services digital network (ISDN), but the prospects for videophony on other networks such as the public switched telephone network (PSTN), private wideband customer-premises networks, cable-television networks, local area networks (LANs) and the future broadband ISDN are also discussed.

5.2. GENERAL DESIGN CONSIDERATIONS

All videophone systems incorporate a camera and a display, and hence have similar problems of obtaining a usable input picture, despite camera siting constraints and variable lighting conditions, and of displaying images of a suitable size in as compact an equipment as possible.

5.2.1 Display

The optimum size of the display screen in a videophone terminal depends on a number of factors: the picture scanning standard, picture resolution (signal bandwidth), viewing distance and, if a data compression scheme is implemented, the visibility of coding artefacts. The 625-line television scanning standard in Europe (525-line in the USA and Japan) was chosen so that the line structure would be imperceptible to the majority of people when viewed at a distance greater than six times the picture height (a viewing ratio of 6H).

Videophone terminals normally use 625-line or 525-line displays, and are positioned on a desk or table between 60 and 120 cm from the user. To enable controls to be easily reached a separation of about 70 cm is appropriate, at which distance the display size should be about 20 cm diagonal to preserve a viewing ratio of 6H. In practice, designers of wideband terminals may opt for a larger display (25-35 cm diagonal), preferring the greater impact of the bigger picture despite some visibility of the line structure. Equally, where low bit-rate coding algorithms are used, usually accompanied by a deliberate reduction of the available resolution, designers may prefer a smaller size (10-15 cm) to mask the loss of picture quality incurred.

The most important advance in display technology over the past few years has been the development of colour liquid crystal displays (LCD) with the grey-scale resolution required for television [1]. Production LCDs are now available in sizes up to at least 15 cm diagonal and virtually all employ an active matrix of thin-film transistors to control the transmittance of each cell. Colour is provided by having three cells per picture element and an overlay colour filter. A light source must be provided behind the LCD, usually by means of a colour-matched cold-cathode fluorescent tube. The main attraction of the LCD is that there is much more freedom to design an attractive, compact unit than would be the case with conventional cathode-ray tube displays; charged-couple device camera sensors have the same advantage over camera tubes. A demonstration videophone terminal incorporating a 15 cm LCD and a CCD camera is illustrated in Fig. 5.1.

Fig. 5.1 Demonstration desk-top videophone terminal.

An attractive alternative may be to integrate the videophone facility into a personal computer (PC), particularly where the potential user already has such a unit on his desk. In this case, where low bit-rate coding is utilized, a reasonable implementation is to 'window' the small videophone picture into one of the top corners of the PC screen as illustrated in Fig. 5.2. For wideband videophone terminals, the whole PC screen can be used for the displayed picture if desired.

88 VIDEOPHONY

Fig. 5.2   Videophony integrated into personal computer.

5.2.2   Camera

Assuming a desktop terminal with a nominal viewing distance of about 70 cm, a camera incorporated in the terminal should have a fixed field of view of about 44° horizontally and 33° vertically, corresponding to a width of field in the subject plane of about 560 mm. With an industry standard 0.5-in format camera sensor (active area 6.25 mm by 4.75 mm), the focal length of the lens should be in the range 7 mm to 8.5 mm.

For most videophone terminals a simple fixed-focus (e.g. at 70 cm) system is perfectly adequate. Even with the lens aperture open at f2.0, the subject will be in focus over the range 45 cm to 1.5 m for a medium-resolution picture (256 picture elements per line), and over the range 55 cm to 95 cm for a high-resolution picture (512 picture elements per line).

Videophone terminals need to operate satisfactorily in lighting conditions which range from bright sunlight to poor artificial lighting. It is the illumination of the user's face that is important, and the camera sensitivity, auto-iris lens performance and electronic automatic gain control must be such that satisfactory pictures are obtained over a range of illumination of 200 to 4000 lux. In addition, the colour temperature varies among different types of artificial light and the colour temperature of daylight also varies with time of day and prevailing weather conditions. Therefore the videophone camera needs an automatic (or at least semi-automatic) means of controlling the colour balance in the television pictures so that, for example, white objects always appear white and even more importantly, facial flesh tones are well reproduced. The range of colour temperature over which such colour balance must be maintained is typically from 3000 to 6000 K.

Another important aspect of the camera specification is that the field rate of the camera should be the same as the local mains frequency; when cameras are operated in artificial lighting, particularly from fluorescent tubes, the video signal is modulated by the illumination frequency (twice mains frequency). If, for example, a 60 field/s camera is operated in 50 Hz lighting, there is 20 Hz modulation flicker running through the displayed picture and 20 Hz is close to the most visible temporal input frequency for the human eye – brain system. Another disadvantage of such a system is that if a conditional replenishment coding scheme (such as that described in CCITT Recommendation H.261 [2]) which encodes differences between the current and previous pictures is used, this modulation flicker will be interpreted as a difference signal and valuable transmission capacity will be used unnecessarily to encode the flicker. Means for removing or at least reducing the flicker exist, but the expense and complexity involved probably outweigh the advantage of having a single worldwide videophone camera and display standard.

To provide the most natural conditions for a videophone conversation, good eye contact with the displayed image of the remote user is essential and this implies that the ideal position for the camera is near the centre of the display screen. Systems using a semi-silvered mirror as shown in Fig. 5.3(a) to achieve this apparent position have been demonstrated, although there is an inevitable consequential light loss of a factor of two to the camera and from the display. When the camera is mounted in the plane of the display, the options are to position the camera centrally above or below the display (see Fig. 5.3(b)) or on one side of the display on the anticipated eyeline of the displayed head-and-shoulders picture, i.e. slightly above the centre line as shown in Fig. 5.3(c). Figures 5.3(b) and 5.3(c) also give the eye contact angles for 10, 20 and 30 cm displays at the same viewing distance of 70 cm, assuming that the centre of the lens cannot be positioned closer than 3 cm to the edge of the display. The advantage, in terms of reduced eye contact angle, of positioning the camera above rather than at the side is immediately obvious. However, despite being a less than optimum arrangement, there are a number of experimental videophone terminals (usually the smaller ones) with the camera positioned at the side. One of the reasons the industrial designers concerned give for preferring this arrangement is that a camera at the side is less obtrusive and dominant than one positioned centrally above the display. Further work is required to determine user preference.

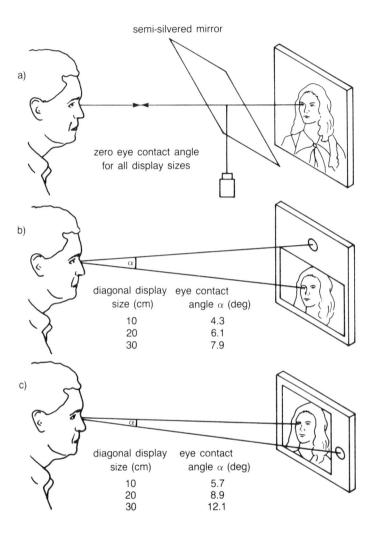

Fig. 5.3 Eye-contact angles for various configurations and display sizes in videophone terminals when the viewing distance is 70 cm: (a) virtual position of camera directly behind display; (b) camera positioned above display; (c) camera positioned at side of display at eye-line.

A European standard is in preparation which addresses these and many other design aspects of videophone terminals, including the control and indication functions required to set up a video call, and standardization of the man−machine interface [3].

5.3. VIDEOPHONES FOR THE ISDN

The ISDN is a customer-switched public digital network. Basic rate access to the ISDN provides two full-duplex 64 bit/s digital connections (the bearer or B-channels) and a full-duplex 16 kbit/s signalling channel (the data or D-channel), and is commonly referred to by the notation 2B+D.

Initially, the marketing of ISDN is likely to be targeted on business users where the demand for a wide range of telecommunications services exists and is growing. The market for videophony among the business community has yet to be established, and its success will depend crucially on the cost of such a new service. If it were possible to replace an existing telephone with a videophone at no extra charge for the terminal or for making calls, many people would probably do so. The technical challenge is therefore to develop a videophone terminal which costs no more than a top-of-the-range telephone, yet includes the additional complexities of a video codec, camera, monitor, audio codec, and full ISDN call set-up facilities.

5.3.1 General design considerations

An important ISDN videophone design consideration is whether the connection should utilize one or both of the B-channels. If single B-channel use is specified, the following advantages apply:

- Call charges are lower.

- The possibility of having to deal with one of the channels being 'busy' is eliminated.

- The need to compensate for different transmission delays (e.g. because one of the channels is routed via satellite and the other via cable on an international connection) is avoided.

However, there are also disadvantages:

- The bit rate available for picture transmission is restricted to about 46 kbit/s, with a consequential reduction in picture quality.

- The expense of a 16 kbit/s audio codec is incurred, and audio quality can then be no better than conventional telephony.

The CCITT has already anticipated that a single 64 kbit/s channel may be required to carry a mix of speech and data simultaneously and has specified how this must be done in Recommendation H.221 [4].

A number of different types of videophone terminal are possible. A stand-alone video terminal and a PC-based terminal with, in each case, ISDN call set-up from a separate telephone, are illustrated in Figs. 5.1 and 5.2 respectively; the advantage here is that users with existing ISDN telephones can continue to use them, though there is no guarantee of harmony of style between the telephone and videophone cases. Another option is to integrate the telephone and videophone within the same case to give a simple, compact unit which minimizes the footprint on the desk and the number of interconnecting cables.

For people to be able to converse freely via a telecommunications medium the transmission delay should be less than a few tens of milliseconds; however, with the scheme specified in Recommendation H.261, it is difficult to design a practical video codec with a delay of less than 300 ms. To give the appearance of lip synchronization, the timing of the sound relative to the picture should be in the range -40 ms to $+20$ ms [5], and although the video delay can vary with the amount of movement to be coded, a fixed compensating audio delay, if chosen carefully, may be acceptable. With around 350 ms delay in each direction of transmission, the ease of conducting an interactive conversation is adversely affected — rather worse than telephony via satellite — but the presence of a picture may help to avoid the tendency for both parties to speak at the same time as often experienced by telephone users when a connection is via satellite. Alternatively users may prefer to minimize the audio delay and accept the loss of lip synchronizm.

A loudspeaking telephone (LST) facility allows the most natural form of videophone communications, avoiding pictures of the remote user with a handset covering a significant part of the face. Given the large round-trip transmission delay, any perceptible echo from the remote end would be extremely annnoying; such an echo must be reduced to around 45 dB below the direct speech level. This is achievable in a good handset, but in LST use, the acoustic path from the loudspeaker to the microphone is always significant; either this acoustic return must be cancelled electronically, or a voice-activated variable attenuator to suppress the echo must be included. The problem is exacerbated by the reduced stability margin resulting from the long echo delay time.

One further problem associated with digital telephony is that different companding laws (A-law in Europe, μ-law in Japan and the USA) are used.

For international digital telephony, A-law/μ-law conversion takes place at the 'gateway' telephone exchange. For international videophone calls using a single 64 kbit/s channel, an 'unrestricted digital' connection is required, but there is no problem because 16 kbit/s audio must be used. A problem does arise, however, where, having specified 'unrestricted digital' transmission for an international call to a region with different standards, the remote terminal happens to be a telephone and not a videophone. In this case, the videophone must recognize the incoming coding law and adopt it. Alternatively it may re-dial specifying 'speech' so that the call is routed via the companding law converters at the gateway.

5.3.2 Hardware design

An ISDN videophone using one B-channel must include all the functionality of a voice-only ISDN telephone, and also incorporate the additional hardware required for the following:

- camera, monitor, and associated control circuitry;
- 48 kbit/s video codec, including error correction;
- 16 kbit/s audio codec;
- hands-free speech, microphone/loudspeaker (if required);
- multiplexing of audio and video.

A major part of the design problem for a videophone lies with the video codec. The magnitude of the data compression required to encode moving colour pictures for transmission at 64 kbit/s is apparent when it is considered that the digital standards for television studios [6] result in a gross bit rate of 216 Mbit/s. On this basis, the compression ratio is over 3000 to 1. However, by encoding only the active picture area, reducing the picture resolution to 176×144 luminance pixels (each colour component 88×72 pixels), and dropping the frame rate to 10 picture/s, the coding compression ratio falls to less than 50 to 1: still a non-trivial coding task but one which has the potential for a cheap and compact implementation appropriate for videophones.

Since video codecs for use on the ISDN only need to process data for transmission at 64 kbit/s, there is a choice for the architecture of the codec: either a conventional hardware implementation or, given the processing power of recently available digital signal processors (DSP), one in which the source coding algorithm is implemented entirely in software. The latter, in addition to giving more flexibility at the development stage, also gives a very compact

94 VIDEOPHONY

implementation that could only be matched by a hardware structure if there were an accompanying large investment in VLSI. Using state-of-the-art general purposes DSPs in conjunction with custom 'application specific integrated circuits' (ASICs), a 64 kbit/s video codec on a single printed circuit card has been implemented (see Fig. 5.4). More details about both types of video codec architecture — hardware and software — are given in Chapter 8.

Fig. 5.4   64 kbit/s video codec on a single printed circuit card.

The audio coding is performed by a single DSP and some simple interface circuitry. The other hardware requirements such as LST, multiplexing and error control require further use of ASICs and DSPs. The power consumption of all the above certainly exceed the limits of line powering, so an additional requirement is a mains power supply capable of being activated automatically when a call is being set up or received.

As noted above, the acceptability of videophones is likely to be strongly dependent on cost, and the main barrier to market penetration is initially the difficulty of building all the above hardware into a compact desktop unit at a low enough price.

5.4.   VIDEOPHONES FOR OTHER NETWORKS

Although the ISDN is considered the most suitable available network to support an emerging videophone service over the next five years, it is worth considering the ability of other existing and future networks to support a videophone service, and the consequential effects on the service.

5.4.1 Broadband networks

If customer-switched digital broadband networks were widely available, and call charges were acceptably low, the main advantages for a videophone service would be that the picture quality would be higher, the processing delay would be small and the video codec would be simpler. There is still some flexibility in matching the service to the available bit rate by defining the spatial and temporal resolution of the videophones accordingly: as Table 5.1 illustrates, a quarter CIF resolution at 20 pictures/s requires transmission at 3 Mbit/s, using simple pulse code modulation (PCM) whereas broadcast picture quality at 25 pictures/s would require about 125 Mbit/s. There is currently much activity in the CCITT defining the broadband ISDN (B-ISDN), with a broadband user-network interface rate of about 155 Mbit/s, of which up to about 130 Mbit/s is available for customer data. The optimum rate for videophony will depend on the tariffing structure adopted for B-ISDN, but in any case, given the long lead-time to plan the introduction of such a major change to the access network, it seems unlikely that B-ISDN will become widely available for some years to come.

Table 5.1 Required bit rates for the transmission of television pictures of various resolutions using simple PCM

Application	Spatial resolution		Temporal resolution pictures/s	Bit rate Mbit/s
	Luminance signal	Each colour difference signal		
broadcast television	720 × 576	360 × 288	25	124.4
video-conferencing (CIF)	352 × 288 352 × 288 352 × 288	176 × 144 176 × 144 176 × 144	30 20 10	36.5 24.3 12.2
videophony (¼ CIF)	176 × 144 176 × 144 176 × 144	88 × 72 88 × 72 88 × 72	20 10 5	6.1 3.0 1.5

However, there is a greater likelihood that broadband customer-premises networks or closed user-group networks (switched cable TV networks with broadband return channel) will become available sooner. Since the networks are specific to closed groups of users, the option of wideband analogue transmission exists, but as more use is made of optical fibres and as digital

96 VIDEOPHONY

electronics technology continues to increase in complexity while reducing in size and cost, the digital option becomes increasingly attractive. The big disadvantage of broadband networks is that they are essentially service-specific (certainly so far as both-way video transmission is concerned), and the cost of providing full-duplex transmission and switching has therefore to be justified on the basis of a videophone or videoconference service alone, until such time as other services requiring high bit-rate full-duplex transmission emerge.

5.4.2   LANs

A local area network, principally used for interconnecting PCs and workstations, is one type of customer-premises digital network that may be capable of supporting a videophone service. The data to be transmitted over a LAN is assembled into packets for transmission, and the ability of a terminal to transmit a particular packet depends on a number of factors including how many other terminals are trying to do likewise at the same time; the transmission is therefore essentially asynchronous with an unspecified transmission delay. While it does not normally matter whether a computer file is transferred in a few seconds or a few tenths of a second, the quality of real-time services such as speech and videophony would be severely affected unless an almost fairly steady throughput and bounded delay could be guaranteed. For this reason, LANs such as Ethernet in common use today are not suitable for real-time services; however, the next generation of LANs will be based on the token-exchange principle and are likely to be much faster and more intelligent (able to prioritize data), and so could well support videophone services of an acceptable quality.

5.4.3   PSTN

So far, we have discussed videophony at 64 kbit/s and over high bit-rate networks and LANs. Could the existing public switched telephone network (the PSTN), which has the advantage of being the most widespread network available, possibly support a videophone service?

The PSTN is designed to carry speech in an analogue bandwidth of around 3 kHz, over a thousand times less than that required for an analogue television signal. It is theoretically possible to devise a new television standard having a resolution low enough to be accommodated in 3 kHz: for example a picture of only 25 lines, with 25 pixels per line and a frame rate of 10 frame/s — this would in fact be similar to Baird's first pictures! However, the severe phase distortion at the band edges of a PSTN channel, plus the overheads of television synchronization pulses, would reduce the usable bandwidth even further to around 2 kHz. Even then, a second telephone channel would be needed for the accompanying speech.

An alternative approach is to encode the pictures and speech digitally and multiplex the data streams for transmission via a data modem. Currently available full-duplex PSTN modems can operate at a rate of 14.4 kbit/s, and designers are confident that 19.2 kbit/s will be possible in the near future; 19.2 kbit/s modems for use on four-wire circuits are already available commercially. In 14.4 kbit/s operation, the speech data would need to be encoded to about 4.8 kbit/s and the picture data coded to about 9.6 kbit/s. On the assumption that a video codec can compress the video data by a factor of 50 (similar to the compression achievable with existing 64 kbit/s codecs), then a colour picture of resolution 88×72 (luminance), transmitted at 9.6 kbit/s, should result in about 6 pictures/s; for monochrome pictures only, the rate increases to about 9 pictures/s. So a videophone service based on small, low-resolution pictures is not out of the question but, as with videophony over the ISDN, the crucial factor is the cost of the complete service, given the speech and picture quality provided. The complexity of a PSTN videophone terminal is greater than that for an ISDN terminal since there is the added cost of a 14.4 kbit/s PSTN modem. The only hope to reduce the cost significantly is for a manufacturer to be sufficiently convinced of the prospects for large-volume sales to enable him to recover the substantial investment in VLSI necessary.

The above approach has been taken to devise an experimental PSTN video terminal to enable deaf people to communicate effectively using sign language [7], despite the limitations of picture resolution and rate. Unlike conventional videophones, an associated speech channel is not required when the users are proficient with sign language (if the pictures are sufficiently good to allow lip reading however, an accompanying speech channel, perhaps even of reduced quality, would be beneficial since it is known that most lip readers have residual hearing and make use of the accompanying audio cues). In addition, a high picture rate (more than 10 pictures/s) is important, even if achieved at the expense of poorer spatial resolution.

Earlier versions of the experimental terminal used cartoon (binary) pictures, generated using an algorithm developed at the University of Essex [8], to reduce the amount of picture data required to be transmitted to a minimum. Although deaf people were able to communicate effectively via the cartoon images (see Fig. 5.5), recent advances in image processing have enabled normal grey-scale pictures to be generated at a sufficiently high picture rate, and deaf users have found such pictures to be subjectively more acceptable. Other features available on the experimental terminal will include text transmission (compatible with existing text transmission terminals),

98 VIDEOPHONY

Fig. 5.5 An experimental PSTN video terminal to enable deaf people to communicate using sign language.

continuous self-view facility to enable a user to remain the field of view of the camera, a user-friendly terminal interface (visual indications of call progress and status normally indicated to the user by telephony tones) and, on some terminals, optional low bit-rate audio. The results of trials of this equipment by volunteer deaf people will be available in mid-1992.

5.5. CONCLUSIONS

This chapter has reviewed the prospects of providing a videophone service over the ISDN, as well as a number of other existing and future networks such as switched broadband networks, LANs and the PSTN. In addition, some of the design aspects for videophone terminals have been considered.

Owing to simultaneous advances in signal processing technology, and the increasing availability of high-capacity switched networks, the age of the videophone is about to begin. In the next few years, videophone products will be extensively trialled on pilot ISDNs in many countries.

Whether or not a videophone service is successfully introduced depends on its perceived utility given the cost of the service and the terminal equipment. The first telephones were regarded as an intrusion of privacy, but the advantage of ease of communication soon overcame that hurdle. Much human interaction is non-verbal, and it will be interesting to see if the videophone also wins acceptance as something which improves the quality of personal communication.

One further factor to bear in mind is that, once introduced, the quality of a videophone service will continue to improve as further advances in image and speech processing are made, and the cost of the service and terminal equipment will continue to fall as advances in integrated circuit technology enable more complex processing to be implemented into smaller but faster chips. Both factors suggest that the growth of the service will follow classical trends and it is anticipated that the rapid increase in growth will take place in the mid 1990s.

REFERENCES

1. White J C: 'Colour LCD TV', Phys Technol, $\underline{19}$ (1988).

2. CCITT Recommendation H.261: 'Video codec for audiovisual services at $p \times 64$ kbit/s', Geneva (1990).

3. European Telecommunications Standards Institute provisional Standard ETS 300 145: 'ISDN and other telecommunications networks — audiovisual services — narrowband visual telephone system' (1991).

4. CCITT Recommendation H.221: 'Frame structure for a 64 to 1920 kbit/s channel in audiovisual teleservices', Geneva (1990).

5. CCIR new Recommendation 717: 'Tolerances for transmission time differences between the vision and sound components of a television signal', Dusseldorf (1990).

6. CCIR Recommendation 601: 'Encoding parameters of digital television for studios', Geneva (1986).

7. Whybray M W: 'Visual telecomms for the deaf at 14.4 kbit/s on the PSTN', COST 219 Videophony for the Handicapped Seminar, The Hague (1991).

8. Whybray M W and Hanna E: 'A DSP based videophone for the hearing impaired using valledge processed pictures', IEEE Conference on Acoustics, Speech, and Signal Processing, pp 1866—1869, Glasgow (May 1989).

6

MULTIPOINT AUDIOVISUAL TELECOMMUNICATIONS

W J Clark, B Lee, D E Lewis and T I Mason

6.1. INTRODUCTION

In general, telecommunications systems have been designed to operate on a point-to-point basis, between two end-users or between a user and a centralized facility such as a database. Where communication is needed between a number of locations, for instance facsimile transmission between a head office and a number of branch offices, then this has been achieved either by means of a store-and-forward facility or by sequential transmission from one site to each of the others. Whilst such techniques may be adequate for non-real-time applications, in cases where human communication is sought by voice or picture, simultaneous reception of information by a number of sites is necessary. Multipoint services provide for this real-time transmission between three or more locations. The generic term 'teleconferencing' is also often used, though this also covers the case where a number of people are involved on a purely point-to-point link. An increasing number of businesses operate on a multisite basis, and the need for standardized multipoint services is apparent. This chapter describes the history and development of various types of multipoint service.

6.2. HISTORY

From the earliest days of the telephone, the opportunity has been taken to provide additional services. The first use of what might be called multipoint telephony took place during the last century when the sound of a live performance from the opera stage was sent to a number of subscribers. In this system, the sound was transmitted in one direction only and owing to the lack of amplification only a limited number of receivers could be used. The first multipoint audioconferences took place in the 1930s, linking housebound and hospitalized students in a school district of Iowa [1].

In a simple audioconference, a number of individuals, each in a separate location, can confer using normal telephone instruments connected via the public switched telephone network to a special telephone conferencing bridge. To overcome the fatigue of using the handset for long periods, loudspeaking telephones (LSTs) were developed which provided the user with hands-free facilities, using a voice-switching technique for half-duplex operation to overcome the problem of acoustic feedback from loudspeaker to microphone [2].

To overcome some of the inherent deficiencies of simple voice-switched LSTs, systems such as ORATOR [3] were developed; this used two separate telephone lines and a specially designed loudspeaker and microphone combination. Using microprocessor-controlled electronics within the terminal to overcome some of the problems of line loss / frequency response and room characteristics, ORATOR provided a convenient audioconferencing system on the public telephone network.

The systems described so far have been solely audioconferencing systems. Once the idea of linking a number of sites for audio had been established, it was natural to attempt to add some form of pictorial or graphical information. Basically two types of additional information can be considered: still images such as facsimile and still-picture television—(SPTV); and moving video images. The combination of these information sources with audio are known as audiographic teleconference and videoconference respectively. For historical reasons these have tended to be considered as separate services, although as described in section 9, one can envisage systems providing a single service across a whole range of facilities.

Audiographic teleconferencing started in the late 1960s using the telephone network to provide audio together with some supplementary facility. Early examples of this are the use of audio and facsimile by NASA to co-ordinate the Apollo program, and the use of a mechanical telewriter with audio for mathematics tuition. In the late 1970s, various experiments were conducted into the use of SPTV in conjunction with audio. In 1980, BT Laboratories implemented trials [4] of applications for SPTV, including telemedicine and teleconference. Resulting from this experience, a successful system was developed known as IMTRAN [5] which provided for the transmission of body scanner images from a hospital to a remote consultant.

In 1981, BT Laboratories and the Open University conducted a human-factors trial of the CYCLOPS [6] system. This equipment combined a loudspeaking telephone terminal with the means to exchange text, drawing and simple images displayed on a television screen; a light pen enabled modification of the displays and thus an additional means of interaction between lecturer and students. Interconnected by means of telephone bridges, some 16 terminals were installed in study centres situated in the East Midlands, and valuable experience of operating such a system was obtained.

Although videoconference uses larger communications resources and more equipment than audiographic conferencing, it has the longer history. Confravision [7] opened in 1971 as a public service between London, Manchester, Birmingham, Bristol and Glasgow. By 1990 several thousand public and private videoconference rooms were operational in the world. Although at its commencement videoconferencing used purely analogue techniques, the key impetus to its development was given by the advances in video coding [8] which together with digital transmission enabled an economical service to be provided.

When the first video codecs were developed in the early 1980s, consideration was given to the need for multipoint videoconference working; a multipoint control unit (MCU) was developed to control the signal flows between terminals at several different locations, to form a multi-site conferencing system, further described in section 6.5.

6.3. BASIC NETWORK CONFIGURATIONS

The objective of multipoint teleconferencing is to enable a number of participants at remote locations to communicate in as natural a way as possible; the system should be easy to use and economical.

Existing equipment and interconnection networks provide primarily for point-to-point connections; a multipoint conference system must be capable of working with such equipment. Within this constraint two network topologies, star and mesh, naturally suggest themselves.

The mesh network (Fig. 6.1) allows a connection from each individual site to all other participating sites and can be thought of as a number of point-to-point links. However, there is a large disadvantage in that as the number of locations requiring connection increases the number of point-to-point connections grows more rapidly. As an example, a multipoint conference between three locations requires only three bidirectional links, but a multipoint conference between six locations requires 15. Also the amount of equipment provided at each location increases commensurately.

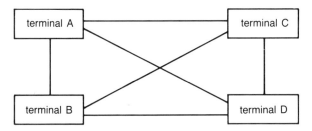

Fig. 6.1   Typical mesh network.

The star network allows each location in a multipoint conference to be connected to a central node. From this node, information may be selected, processed and distributed to all other points of the star, thus allowing for a reduction in required bandwidth relative to the mesh type network. Also it is economical in the amount of equipment to be provided at the participant locations. However, the star configuration requires extra equipment in the form of an MCU, located at the hub of the star as shown in Fig. 6.2. The star network may be easily expanded into a double star or 'dumbell' configuration in which one point of each star network is connected to one point of the other, as shown in Fig. 6.3. This is quite a useful configuration: an MCU may act as a concentrator for several sites communicating via a single bidirectional link with several other sites connected to a second MCU, perhaps in another country.

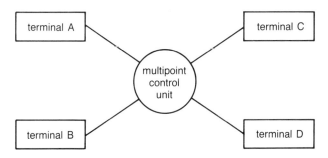

Fig. 6.2   Typical star network.

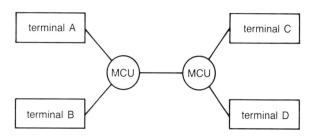

Fig. 6.3 Typical dumb-bell network

6.4. TYPES OF MULTIPOINT CONFERENCE

The ideal multipoint conference would provide every location with access to all available data, speech and video; this is known as 'continuous-presence' multipoint conferencing for which the mesh type of network is ideally suited.

An alternative to continuous-presence multipoint conferencing is switched multipoint conferencing, whereby video and data information is selected by some criteria and distributed to connected locations. The selection criterion may be fully automatic, i.e. voice activated, or may be manually controlled, perhaps by a chairperson. Audio is mixed at the MCU in such a way that any particular location does not receive its own audio but receives the sum of all others.

6.5. VIDEOCONFERENCE MCU

Digital videoconference codecs were first standardized in 1984 as CCITT Recommendations H.120/130, for transmission on a 2 Mbit/s or 1.5 Mbit/s link, and a multipoint system was needed to interconnect three or more of these. A star network with switched multipoint conferencing [9] was chosen as the most suitable configuration, requiring the development of an MCU. Similar considerations applied to the lower bit-rate videoconferencing systems based on the Recommendation H.320 adopted in 1990.

An MCU has a number of ports, each communicating with a single videoconference codec. The compressed video signal is switched according to conferees' voice activity, the audio level being monitored by the MCU. Thus, in Fig. 6.4, the video from terminal B is identified as the 'current speaker' and distributed to all other locations. The picture received by the speaker is that of terminal C, the 'previous speaker'. This allows the conference to proceed in a fully automatic mode.

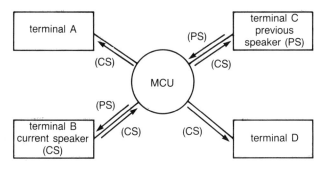

Fig. 6.4 Video switching technique.

The MCU may be considered as an audio conference bridge with a switched video overlay [10]. A prototype MCU was developed in 1983 using a four-port switch capable of 2 Mbit/s operation only. Field trials were conducted with European PTTs using a satellite to provide interconnections, the first time compressed digital videoconference signals had been demonstrated in multipoint configuration; a number of standardization working party meetings were held using the system to test its usefulness in a real-life meeting situation. The system was subsequently adopted by CCITT as Recommendation H.140.

6.6. NEED FOR CONTROL AND INDICATIONS

The systems described so far have been automatic in operation; that is, once the conference has been set up, all participants have equal status and any switching of the audio or video signal is carried out by the equipment itself without participant intervention. Face-to-face meetings tend to fall into two categories: they may be very formally structured, with participants having to gain permission to speak from a chairperson; or they may be small, informal meetings where all participants have equal status. These two types of conference may be called 'conducted' or 'non-conducted' respectively. In a conducted conference, the chairperson may speak and be heard by other participants at any time. However, if a participant wishes to make a contribution he must first send a 'request for the floor' message to the chairperson, asking for permission to speak. As in a face-to-face meeting, the chairperson may permit or disallow the contribution.

For a non-conducted conference, no chairperson is necessary and the sound from each terminal is sent to all others. In this case it is left to the participants to moderate their speech so that a meaningful discussion can be held.

For some types of meeting it is desirable that means be provided for a chairperson to control the conference in various ways. For simple audio and video conferences, a manual override on the multipoint switching may be sufficient to control the conference. However, where more complex control of audio or video is desired, or where additional graphical aids are in use, a teleconferencing protocol must be used as described in section 6.7. Such a scheme can also be used to improve the human interface, by providing displayed information about the conference participants, available auxiliary equipment, etc.

6.7. THE MIAC PROJECT

In order to generate and prove some of the protocols mentioned above, the MIAC (multipoint interactive audiovisual communication) collaborative project [11] was set up in January 1986 under the auspices of the ESPRIT programme of the European Community.

There were three main aims to this project:

- to standardize the way in which speech and data signals could be combined and controlled for transmission over single 64 kbit/s digital lines between a number of locations;
- to demonstrate these standards working in an international multipoint audioconferencing system;
- to study the human factors aspects of multipoint multifacility services, by testing the equipment by engineers, human factors experts and naive users.

The two main elements of the MIAC system were the audio conference terminal (ACT), and the MCU. The demonstration equipment produced during the project enabled up to four ACTs to be connected to an MCU; two MCUs could be connected in tandem to form a conference of eight terminals. The ACT consisted of microphones, loudspeaker, and a personal computer running the conferencing software with associated transmission and control cards.

Through MIAC it was possible for the first time to participate in audioconferences with high-quality wideband (7 kHz) speech between many remote locations, with the added facilities of full chairperson control messaging, Group 3 facsimile and SPTV, together with a telewriter.

Extensive demonstrations of the equipment have highlighted what can be expected from the next generation of audioconference services. In the human factors arena, multipoint multifacility conferencing was shown to be a practical and usable service. Figure 6.5 shows the system in use.

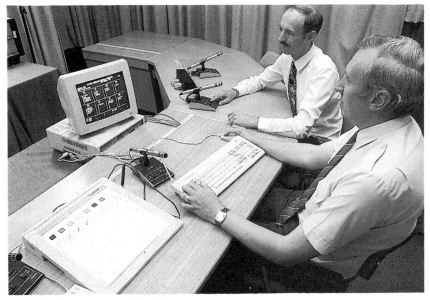

Fig. 6.5   The MIAC system in use.

The set of protocols included a range of about 20 conference enhancements which can be grouped under the following headings:

- conference set-up and clear-down;

- chairperson control of the conference;

- user controls and indications;

- the multipoint operation of meeting aids, e.g. facsimile, SPTV.

Each conference facility is supported by an associated repertoire of control messages, for negotiation of services or handshaking between meeting aids. For instance, Group 3 facsimile is made to operate in a broadcast configuration by connecting it to an interface and protocol adaptor (IPA) in the ACT; the communication between facsimile machine and IPA is effectively point-to-point but that between IPAs is effected by multipoint messages, the IPAs in turn handshaking with their respective facsimile machines; the sending facsimile machine subsequently transmits its data to the others via a transparent data channel, assigned dynamically as required. A similar procedure is followed for other meeting aids, such as SPTV. The control messages are transported in a 4 kbit/s message subchannel and switched at the MCU.

MIAC has adopted the concept of distributed conference control, with conference facility protocols residing in each terminal rather than in the central MCU. In this way a terminal can operate directly to another terminal in a point-to-point mode, or via an MCU in a multipoint mode, with identical conference facilities being available in each case. This approach requires a mechanism by which control information can be routed from any point to any or all other points in the conference network; this routeing is achieved by an address field allocated in the message header.

The security of the conference protocol relies on a secure link protocol, provided by CCITT X.25 LapB on each message channel link between a terminal and the central MCU; the MCU itself is assumed to be error-free. Since each control message in the system is sequence-independent and recovery from error or fault situations is achieved in the applications layer, very lightweight transport protocols suffice. In fact, if compared with the OSI model, layer 3 is a simple 3 byte routeing address and layers 4 and 5 are not used. Since an indicator-length-content structure was originally specified for MIAC messages, each message was formatted according to CCITT Recommendation X.409, this being the equivalent of a layer 6, with the individual facilities forming the layer 7 applications.

The conference protocol developed was adopted in European Telecommunications Standard (ETS) T/N31-02 concerned with audiographic teleconferencing and the associated message channel as ETS T/N34-02.

The audio, message channel, and meeting-aid signals were multiplexed using a 64 kbit/s frame structure which offered the following features:

- frame alignment and bit rate allocation signals (FAS, BAS);

- audio 56 kbit/s (optionally 48 kbit/s) to G.722;

- message channel 4 kbit/s;

- transparent data channel 8 kbit/s (optional);

- telewriter channel 800 bit/s.

This frame structure was subsequently further developed to become CCITT Recommendation H.221 (Chapter 7).

6.8. THE MIAS PROJECT

Building on the advances made in MIAC, a follow-on ESPRIT project known as MIAS (multipoint interactive audiovisual system) was carried out in the period 1988-91. MIAC had shown some of the possibilities for the use of a single 64 kbit/s channel in an interactive system. It was the intention from the outset that this concept be extendible to multiple 64 kbit/s, so one part of the MIAS project was to implement this option for the new generation of moving-video systems, first at 2×64 kbit/s but with the possibility for extension to higher multiples up to 30×64 kbit/s. Network interfaces to CCITT Recommendation I.420 provide for use over the ISDN.

Within an office environment, a number of different tasks need to be carried out, hitherto usually performed on separate equipments — for instance a facsimile machine cannot be connected directly to a word processor. Increasingly users need to have available a whole range of communications facilities within their own office. By integrating the more usual paper-based facilities such as facsimile with pictures and text on a common screen, a complete workstation can be created which enables many tasks to be simplified. In the MIAS system the infrastructure developed will allow these tasks to be carried out simultaneously between a number of users.

Based on the above concept, the project had the following essentials:

- a proven infrastructure for multimedia audiovisual communications at $p \times 64$ kbit/s ($p = 1$ to 30);

- the development of demonstrator audiovisual terminals for various applications — office-based multimedia terminals providing a variety of facilities based on quality audio including moving pictures, Group 4 facsimile, file transfer and group document editing;

- a multipoint unit compatible with CCITT I-series Recommendations capable of being incorporated either in concept or physically into private or public switching systems;

- a European demonstrator system with terminals sited in partners' premises.

In addition to the features of MIAC, using the second 64 kbit/s channel it is possible to provide moving video or higher-speed meeting aids such as

Group 4 facsimile, data files, still pictures. As a further enhancement, meeting aid data can be transmitted not only in broadcast mode but addressed to other terminals selectively.

Both the meeting aid data and control messages are now packetized in the message channel; the message channel itself may be expanded in discrete steps from a nominal 6.4 kbit/s data rate up to a maximum of 68.8 kbit/s. The multiplexed data stream in the message channel contains data of different types and priorities. These are new requirements compared to those met by the earlier MIAC transport protocol:

- priority mechanism;
- sequence control for bulk data;
- more robust error protection;
- selective addressing of data.

These have been incorporated by enhancing the MIAC protocol to provide a number of simultaneous sessions, whose relative priority is determined by additional information in the message routing field. A priority mechanism is necessary to ensure that time-critical interactive data is not delayed by the simultaneous transmission of bulk data such as facsimile. Figure 6.6 shows the protocol structure [12].

for future extensions								layer
			ft	fax 4	point	edit	control	7 application
null	null	null	null	null	x208	x208	x208	6 coding
			4	3	2	1	0	5 session
null					null			4 transport
non priority					priority			3 routeing
lap b								2 link
4—66.4 kbit/s via muldex								1 physical

Fig. 6.6 The MIAS protocol stack.

Human factors testing [13] of the MIAC system was carried out to assess the acceptability of such systems for potential users and consisted of trial conferences with durations varying between approximately one and two hours. Some of the conferences were point-to-point, the others being multipoint; some were international, within Europe, others were just local. Participants were given real communicative tasks during the test conferences so that a visible purpose to the exercise was apparent.

Human factors experts were present as observers at all the test conferences and each participant filled in a user evaluation questionnaire afterwards. Both the analysis of the questionnaires and the reactions from participants showed that the MIAC system had been very positively regarded. The high quality of the audio was one main reason for the positive assessment.

6.9. CONCLUSIONS

Teleconference in various forms has been available for a number of years, but apart from moving video the other visual aids have been provided on an *ad hoc* basis. With network and terminal improvements, and the availability of standards across the whole range of media, multipoint multifacility services are becoming a reality. The acceptability of such services depends not only on their technical performance, but also on the perceived user benefits.

REFERENCES

1. Olgren C H and Parker L A: 'Teleconferencing Technology and Applications', Apr tech House Inc, 1983.

2. Ryall L E: 'A new subscriber's loudspeaking telephone', POEEJ, 29, p 6 (April 1936).

3. Groves I S: 'ORATOR' — The Post Office Audio Teleconference System', IEE Communications, 80 Conference (16-18 April 1980).

4. Kenyon N D: 'Slow-scan TV goes on trial', Br Telecom J, 1, pt 4, p 2 (1981).

5. Ling J T and Redstall M W: 'IMTRAN image transfer system', Br Telecom Eng J, 5, p 197 (Oct 1986).

6. Clark W J: 'Cyclops - a field evaluation', 1982 International Zurich Seminar on Digital Communications, Zurich, (March 1982).

7. Haworth J E: 'Confravision', POEEJ, 64, p 220 (Jan 1972).

8. Nicol R C and Duffy T S: 'A codec system for worldwide videoconferencing', Professional Video, (Nov 1983).

9. Nagra A: 'Multipoint video teleconferencing using conditional replenishment coding techniques', MSc Report, Essex University.

10. Mason T I and Lewis D E: 'Multipoint videoconferencing', IEE Colloquium, 'Video conferencing, has the time come' (Dec 1987).

11. Clark W J, Lee B and Meijboom A: 'The development and evaluation of a multipoint audiographic conferencing system — MIAC'. Proceedings of the 5th Annual ESPRIT Conference, Brussels, p1345 (1988).

12. Coolegem K G, Clark W J, Ceruti R: 'Multimedia desktop conferencing with MIAS', Proceedings of ISSLS-91 conference, Amsterdam.

13. MIAC Consortium: 'ESPRIT Project 1057 MIAC final technical report' (Sept 1988).

7

INFRASTRUCTURE STANDARDS FOR DIGITAL AUDIOVISUAL SYSTEMS

N D Kenyon

7.1. AUDIOVISUAL SERVICES

7.1.1 General terminal

Figure 7.1 is a schematic representation of a generalized communication terminal distinguishing various types of information representation — sound, 'natural' images (still or moving), document images, text, computer graphics, X-Y (telewriter or cursor), control and other data. The term 'multimedia' tends to be applied to a system utilizing more than one of these simultaneously, particularly combinations which include speech[1].

Recognizing that communication terminals may or may not be attended by a human being, the figure is set out with human input to the left, output to the right, and the centre ground occupied by the unattended machine which at any moment may be input, output, or both. In identifying 'inputs' and 'outputs', it is convenient to focus on the 'electrical-to-other' transducers involved. On this basis the facsimile facility is visual rather than mechanical, and video camera input may be of any kind of image — people, documents,

---

[1] There are two fundamentally different approaches to the development of multimedia systems, one in which a real-time conversational capability is added to computer data in the form of text, still images and other non-real-time formats; the other involving the addition of such computer data to audio- or video-teleconferencing. It will be important to ensure that these approaches converge on the same result.

114 DIGITAL AUDIOVISUAL SYSTEMS

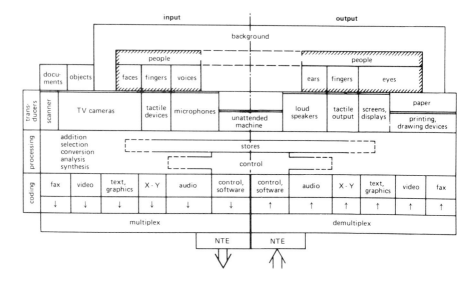

Fig. 7.1 General scheme of audiovisual terminals.

objects, or scenes. It follows that human input may be sound, image, or tactile/mechanical, and output to human beings likewise.

In principle, all the applications mentioned in Chapter 1 can be represented by the generalized terminal of Fig. 7.1. One may note that the layered hierarchical structure has some similarity to that of the protocols under development for open systems interconnection (OSI), though the parallel is by no means exact: the latter was originally conceived for computer communication, with no special attention given to the temporal requirements of real-time audiovisual services and multipoint working. However, it is clear that similar hierarchical principles could be employed with layer protocols adopted from the OSI set where appropriate.

Terminals which have the form of Fig. 7.1, but which differ quite considerably at the input/output level, will nevertheless be able to communicate if the following conditions are met in a standard way, as follows.

- The corresponding communicated signals are one of the defined limited set of formats.

- These signals are multiplexed into the communication channel in a frame or packet structure which retains the identity of the different formats, in variable combinations as required, in such a way as to make maximum use of the available transmission capacity.

- There are procedures for exchange of the signals such as to make the best use of the available facilities, yet at no time does a terminal receive signals it does not have the power to decode.

7.1.2 Harmonization

Although each of the services/applications mentioned in Chapter 1 could be developed independently of the others, this would in the longer term be greatly disadvantageous. Terminals should be developed so that in general they are able to cope with a variety of applications requiring more or less the same facilities; in the absence of harmonization such terminals would become unnecessarily complex, and cumbersome to operate.

The following aspects of harmonization can be identified.

- Terminal design — while many aspects of physical design could and should be left to the individual suppliers of the equipment, some parameters must be specified for satisfactory end-to-end service. This applies particularly to audio conditions (level, background) and video conditions (lighting, contrast, etc): as for telephony, poor design at one terminal can cause the greater inconvenience to the remote party. It is also desirable for controls and indications to be reasonably consistent from one service to another (icons, key responses, tones, etc).

- In-band protocols and signal formats for each of the facilities to be transmitted; multiplexing to form a multimedia channel on a single transmission path, or on parallel paths set up by a single calling procedure; arrangements for interworking between different terminals.

- Network aspects and call control — although it is expected that the networks used for these audiovisual services will be general-purpose networks, the network configuration and equipment will have a bearing on the service that can be provided.

In general, experience with new applications is insufficient to be definitive about the terminal design aspects; it is therefore all the more important to standardize the general transmission format and interworking arrangements, and to ensure that these should be flexible to accommodate the changing patterns of utilization as user experience is gained. The existence of transmission standards will do much to facilitate the introduction of new services on a provisional basis, thus providing the experience upon which further standardization can be based.

7.1.3 A framework for development of standards

Considering the many aspects of services and systems requiring a harmonized approach, the CCITT[2] put together in Recommendation H.200 a framework of target Recommendations, the essence of which is shown in Table 7.1. Not all of these would be brought to fruition within a single study period, but the anticipation of probable future requirements would make the achievement of consistency more practicable.

The framework is in four sections[3]:

- AV.100 series, covering a limited number of specific **services**;
- AV.200 series, providing a standardized set of in-band signals applicable to all audiovisual systems, the **infrastructure**;
- AV.300 series, showing how services of the AV.100 series are realized as engineered **systems**, drawing on the AV.200 'tool-kit', some terminal requirements;
- AV.400 series, providing for **call-control**.

The services and systems will be dealt with in section 7.7, following discussion of the more fundamental infrastructure and call-control matters which occupy the next few sections. The harmonized signal infrastructure covers:

- information signals — the various coding algorithms for audio and video, etc;
- multiplexing — the means for combining them, with other data and control, on to appropriate digital paths (single and multiple connections);
- communication procedures — the means of connecting any two terminals together;
- multipoint working — the means of connecting three or more terminals together.

---

[2] CCITT is part of ITU, the International Telecommunications Union; the work of this body is refined by regional standards bodies, such as the European Telecommunications Standards Institute (ETSI). The output of the CCITT is in the form of Recommendations — see [1].

[3] The designations AV.*xyz* do not represent formal Recommendation numbers, but rather are indicative of a position in the framework; the formal numbers in the F, G, H series are assigned just before final adoption.

Table 7.1 Framework for Recommendations for audiovisual services

	Shortened title	Recom.
AV.100	General Audiovisual Services	
AV.101	Teleconference, general	F.701
AV.110	Audiographic Teleconferencing, general	
AV.111	Audiographic Teleconferencing for ISDN	F.711
AV.112	Audiographic Teleconferencing for B-ISDN	
AV.120	Videophone Services, general	F.720
AV.121	Videophone for ISDN	F.721
AV.122	Videophone for B-ISDN	
AV.130	Videoconference, general	F.730
AV.131	Videoconference for ISDN	
AV.131	Videoconference for B-ISDN	
AV.140	AV interactive (Storage/retrieval) services	
AV.150	'Multimedia' services	
AV.200	General AV Infrastructure	
AV.210	Reference Networks	
AV.220	(Transmission Multiplex Structures)	
AV.221	64-2048 kbit/s Frame Structure	H.221
AV.222	Broadband, for conversational services	
AV.223	Broadband, for distribution services	
AV.230	Frame-synchronous Control & Indications	H.230
AV.231	64-2048 kbit/s Multipoint Control	H.231
AV.232	Broadband Multipoint Control	
AV.233	Privacy System	H.233
AV.240	Communication, principles	H.242
AV.241	System aspects of wideband audio	G.725
AV.242	Communication Procedures, 64-2084 kbit/s	H.242
AV.243	Ditto, multipoint	H.243
AV.244	Communication Procedures (ATM)	
AV.250	(Audio coding)	
AV.251	Narrowband speech in 64 kb/s	G.711
AV.252	Wideband (7 kHz) audio in 64 kbit/s	G.722
AV.253	Audio coding at 24/32 kbit/s	
AV.254	Speech coding at 16 kbit/s	G.728
AV.25x	Audio coding for use on B-ISDN	
AV.260	(Video Coding)	
AV.261	Video codec, up to 1856 kbit/s	H.261
AV.262	Still-picture Television	(ISO)
AV.26x	Video coding for use on B-ISDN	
AV.26y	Video coding for retrieval systems	(ISO)
AV.26z	Video coding for broadcasting	
AV.270	(Protocols)	
AV.271	Audiographic Conferencing	T.agc
AV.272	Generic Conference Control	T.gcc
AV.273	Communication Application Profile	T.avps
AV.274	Multipoint Communications Service	T.mcs
AV.300	(Gen.AV systems/terminals)	
AV.310	Teleconference systems and equipment	
AV.320	Visual telephone systems & equipment for ISDN	H.320
AV.321	Visual telephone systems & equipment for B-ISDN	
AV.330	Equipment for Audiovisual Retrieval systems	
AV.400	(Call control)	
AV.410	Reservation systems	
AV.420	Audiovisual Call Control	H.420

R = Recommendation in 1988, or earlier
A = Accelerated procedure, 1990
N = new or revised Recommendation, 1992
L = Recommendation targeted later than 1992

Attention was for some time focused on the requirements for systems which would operate over ISDN and other digital circuits of fixed bit rate up to 2 Mbit/s. More recently CCITT has progressed towards the standardization of broadband networks based on ATM principles, and the audiovisual infrastructure is being extended to cover such networks: these future targets are included in the framework.

It should be noted that the CCITT is not the only body where relevant standards are generated: the author has included for completeness in Table 7.1 some video coding items which are being dealt with by ISO and CMTT[4].

7.2. INFORMATION SIGNALS

The audio, video and other media have some aspects which are common to the various services, as follows.

Audio

In general only one person speaks at a time, and is heard at one or more other terminals; this could imply that the outgoing audio channels from the other terminals are not required for the time being, but when considering use of that capacity for any other purpose it should be borne in mind that (a) in order not to inhibit verbal interruptions by a listener, the return channel should be available for speech at a few milliseconds notice, and (b) the absence of the background noise from otherwise silent locations may confuse or disturb the speaker.

The quality of transmitted speech is dependent on the available transmission capacity and the complexity of the coding equipment.

- Speech and data signals of a bandwidth of 3 kHz or so can be encoded into 64 or 56 kbit/s by the methods of Recommendation G.711; it should be noted that these methods, known as A-law and μ-law respectively, are incompatible.

- Audio signals with a bandwidth of 7 kHz can be encoded into 64, 56 or 48 kbit/s by the algorithm of Recommendation G.722: this algorithm is quite different from that of G.711, but the system-level Recommendations (G.725 and H.242) stipulate that an equipment which can operate in G.722 mode must also be able to work in both A- and μ-law modes of G.711.

---

[4] ISO — International Standards Organisation; CMTT — a television committee within the ITU.

- A Recommendation for 16 kbit/s encoded speech will be completed in 1992 — here again, equipment which will do this must also be capable of G.711. The performance of this algorithm will be at least as good as that of G.721 for speech/data at 32 kbit/s, so it is not thought necessary to include the latter in the harmonized framework. Formal subjective tests have been made to assess the speech quality of all the audio algorithms.

- A method of encoding 15 kHz sound into 128 kbit/s is being standardized in the ISO.

Video

Here it is essential to distinguish between still-picture television (SPTV) and moving pictures (MPTV). In MPTV the picture contents must be updated at least 10 (preferably 25 or 30) times per second to preserve the illusion of motion. SPTV captures 'frozen' images: even if the system frequently updates a changing scene, each image is presented in its own right with no attempt at continuity of motion.

Figure 7.2 shows the time taken to update a full-screen picture of the given resolution, with different degrees of picture compression ranging from simple PCM to complex transform algorithms. However, a constant data rate is not essential. Under various circumstances it may be preferred to send a single high-quality image quickly (requiring a high data rate), perhaps subsequently pointing to this image using a cursor, or annotating it by means of a telewriter. At the other extreme, a continuous sequence of medium-quality pictures at half-minute intervals needs only a low data rate; for other purposes an intermediate bit rate will be appropriate. Where the scene being transmitted is relatively constant, faster updates are possible by transmitting only the changed portions of the picture.

The ISO has standardized an algorithm set for the encoding of SPTV images [2].

For MPTV the bit rate should be as high as possible, consistent with transmission economics; fundamentally, the rate of information generated is highly variable, but it is not generally practicable to use part of the transmission capacity for other facilities during periods of low movement. The picture quality achieved at various bit rates is not easy to specify, since it is the movement rendition rather than the static picture clarity which is compromised as the bit rate is reduced. The Recommendation H.261 has been prepared to cover all rates up to 1856 kbit/s. This Recommendation differs significantly from the audio Recommendations, in that the algorithm is not totally specified. In essence, it is the transmission format which is set out definitely, but there is considerable scope for flexibility in the way that it is used by different equipments. The encoder must have some limited

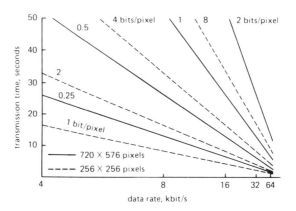

Fig. 7.2 Transmission time for full-screen pictures.

knowledge of the decoder capabilities (see Section 7.5), but within this constraint there is scope for both cost-quality optimization and exercise of the codec designer's skill in arriving at the data to be transmitted. At the decoder end, there is similar scope for optimization of the way in which the incoming data is used to produce the displayed picture. Under these circumstances, standardization of the picture quality is not possible, though test sequences of moving pictures have been chosen for comparison purposes.

ISO has a programme for standardizing algorithms for video coding in the range 1 Mbit/s and above, for applications in a wide range of classes — storage/retrieval, conversational and distribution. CMTT prepares the digital standards used for the transmission of television programmes, notably at 140 Mbit/s and 34 Mbit/s [3,4].

In the future there will undoubtedly be further standardization of MPTV as new image-processing technology becomes available, possibly incorporating model-based techniques (see Chapter 14).

Facsimile

Transmission should be as fast as possible when required, but the facility is not foreseen as being in continuous use during conversational communication. As for SPTV, a constant data rate is not required. There is little experience of the use of facsimile in a multipoint audiovisual environment, but discussion of documents is very much an everyday practice in the business world and must be allowed for in future systems (this applies particularly to 'soft facsimile', where the received pages are displayed on a screen instead of printed out on paper). CCITT has adopted Recommendations T.5, T.6 and T.62, for group 4 facsimile.

Other facilities

Text, graphics, X-Y devices, and control signals typically require only low data rates and although they are probably only used occasionally with audio (people do not often key or draw things while talking) the incorporation of such signals poses no special problem. In addition, data file transfer at a wide range of rates may also find its place in audiovisual applications.

The formats of all such signals are standardized, and it is only necessary, within the harmonized framework, to provide for datapaths suitable to carry them. A number of fixed bit rates must be provided, and also a multilayer protocol (MLP) for telematic signals (see Chapter 6).

7.3. FRAME STRUCTURE FOR MULTIPLEXING INFORMATION FLOWS

Once a connection has been established between the terminals, the transmission capacity available is to be partitioned in such a way as to present the best possible service to the users in a wide range of circumstances, including multipoint working. Flexibility is essential here: the required mix of facilities will vary from application to application, and perhaps also during each call, and the system must accommodate such variations automatically, preferably without the user being aware of them.

Flexible use of a single 64 kbit/s channel may satisfy many applications; for services requiring high visual data rates further 64 kbit/s channels can be added, e.g. a second B-channel as in the basic-access ISDN, or five such channels giving an aggregate 384 kbit/s (the so called H0 channel), or indeed higher multiples of 384 kbit/s.

Where two or more independent connections have been made across the network between the same terminals, it is necessary to synchronize them at the receiver, since the transmission times may not be identical.

As a further complication, in the so-called 'restricted networks' in North America, only 56 kbit/s and multiples thereof will actually be available to the terminal, while at 1536 kbit/s there should never be 16 consecutive zeros. Thus the digital path between terminals may take a wide variety of forms, as listed in Table 7.2.

To multiplex the audio, video and data on to such a path, a frame structure is required with the following properties:

- applicable to the circumstances listed in Table 7.2;
- having means for extracting octet timing;
- having means for synchronizing multiple connections;

- robust to errors, error bursts, slip;
- providing a variable mix of audio, video and data;

and also providing, optionally:

- encryption;
- error-performance checking.

Table 7.2 Digital paths accommodated by H.221/H.242

Transparent networks	Restricted networks
$m \times 64$ kbit/s for $m^* = 1$ to 6	$m \times 56$ kbit/s for $m^* = 1$ to 6
$n \times 384$ kbit/s for $n^* = 1$ to 5	$n \times 336$ kbit/s for $n^* = 1$ to 4
128, 192, 256 kbit/s	112, 168, 224 kbit/s
512, 768, 1152 kbit/s	448, 672, 1008 kbit/s
1536 kbit/s 256	1344 kbit/s
1920 kbit/s 512	

The frame structure having these properties is described in Recommendation H.221, as illustrated in Fig. 7.3. Channels above 64 kbit/s are subdivided into 64 kbit/s time-slots. The structure is octet-based within the first or only 64 kbit/s, each frame having 80 octets, bit 8 being referred to as the service channel (8 kbit/s).

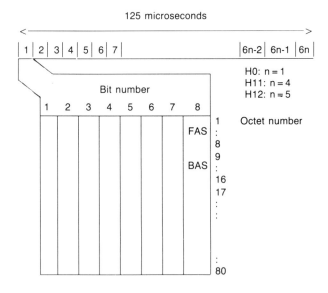

Fig. 7.3   Frame structure.

The first eight bits of the service channel are known as the frame alignment signal (FAS), and perform the following functions:

- frame alignment (hence octet timing if not already available);
- optional multiframe alignment (16 frames);
- optional multiframe counter (modulo 16) for synchronization of multiple connections even if some are via satellite and some terrestrial;
- connection numbering;
- optional cyclic redundancy code checking (CRC4), for error detection on the whole bit stream of the connection;
- the 'A-bit', indicating frame/multiframe/synchronism in the opposite transmission direction.

In a 56 kbit/s connection, the service channel is in bit 7, but there is no other change. In connections of 384, 1536, and 1920 kbit/s the service channel is provided only the designated 'time-slot 1'.

Following the FAS in the service channel, the subsequent eight bits are known as the bit allocation signal (BAS), whose use will be described in section 7.4. If encryption is in use, the initialization vectors are transmitted in the eight bits of the service channel following BAS. The remaining capacity of the service channel and of all the other octet/septet positions is used for a variable mixture of audio, video and data. It can be seen that the overhead of the frame structure is low: only 1.6 kbit/s for FAS and BAS in each connection.

When multiple connections are in use, the first to be established provides the so called 'initial channel', and any audio is inserted into the lowest numbered bits of this channel. Subsequent 'additional channels' are synchronized by buffering at the receiver, and when this is achieved the total capacity is effectively available as if it were a single channel at the combined bit rate. The bit-occupancy of audio and any data is defined by BAS; if video is present it occupies all the remaining capacity, but if there is no video this remains empty.

7.4. COMMUNICATION PROCEDURES

7.4.1 BAS codes

The multiplex structure provides for a wide range of transmitted signal conditions, but it is necessary to follow certain procedures to ensure successful communication between two or more terminals which are not necessarily identical. The procedures (set out in Recommendation H.242) are based on the use of the 8-bit BAS codes, which provide three distinct functions:

- capability codes giving clear indications as to the ability of a terminal to receive and decode the various possible transmitted signals;
- command codes specifying the exact contants of each transmitted frame — they command the demultiplexer at the receiver to operate in a particular way;
- escape codes providing for extension to 16 bits and more, for a variety of purposes.

The values currently defined are listed in Tables 7.3 and 7.4. Once a BAS code has been transmitted, it remains valid until altered by the transmission of an alternative value (that is, a value appearing on the same row in Table 7.4). For example, once the 16 kbit/s speech BAS command has been transmitted, it remains in force until a different audio command is sent.

Table 7.3 Terminal capabilities.

Audio:	audio to Recs G.711 (PCM, A- and μ-laws, G.722 (7 kHz audio, at 64/56/48 kbit/s, 16 kbit/s audio...)
Video:	quarter- and full-CIF pictures to Rec H.261, four values of frame dropping
Transfer rate:	essentially the values listed in Table 7.2
Restricted network:	identifies a terminal connected to such a network
Low-speed data (LSD):	300, 1200, 4800, 6400, 8000, 9600, 14400 kbit/s, 16, 24, 32, 40, 48, 56, 62.4 kbit/s
High-speed data (HSD):	64, 128, 192, 256, 320, 384, 512, 768, 1152, 1536 kbit/s
Low-speed MLP*:	4 kbit/s; 6.4 kbit/s; variable up to 64 kbit/s
High-speed MLP*:	fixed or variable rates
Encryption:	the ability to cope with an encrypted signal as defined in the encryption control channel
Multiple-byte extension:	the ability to cope with a specified internal message system

*MLP refers to the multilayer protocol to be specified in Rec AV.270 series.

Table 7.4 Commands

Audio:	sets the audio to one of the options in Table 7.3, or switched off
Video:	on/off (format and frame-dropping is set inside the video signal)
Transfer rate:	sets the transmission to one of the rates listed in Table 7.2
Restricted network:	obliges a terminal to transmit only signals that can traverse such a network
Low-speed data (LSD):	sets one of the rates listed in Table 7.3, or switched off
High-speed data (HSD):	sets one of the rates listed in Table 7.3, or switched off
Low-speed MLP:	opens an MLP channel at 4, 6.4, or 62.4 kbit/s, or filling the available space
High-speed MLP:	opens an MLP channel at 62.4 kbit/s or higher
Encryption:	on/off

Clearly the BAS codes must be very error-resilient. To achieve this, they are only transmitted in alternate frames, the intervening frames carrying an 8-bit code capable of correcting two errors. Thus for a random-error rate of $1:10^3$ the probability of a BAS code error is still extremely small. Burst errors and slip are more problematic, and for this reason the valid codes should be retransmitted cyclically when the BAS position is not otherwise in use.

In addition to capabilities and demultiplexing commands, the BAS codes provide a number of other controls and indications, covering such functions as audio muting, 'on-air' indicator, and controls for simple multipoint working; these are defined in H.230.

7.4.2 Initialisation

When two terminals are first interconnected by a clear digital channel, initialization takes place as depicted in Fig. 7.4; the FAS and BAS are transmitted in the service channel, the rest of the first 64 kbit/s being occupied by G.711-encoded audio. Both receivers are continuously searching for frame alignment, and when this is achieved the A-bit is set to zero on the outgoing channel. Only when receiving $A = 0$ can a terminal be sure that the remote terminal can understand and act upon a BAS code.

Receiving $A = 0$, the calling terminal X can then transmit all its capability codes. In due course it will also receive a set of capability codes from terminal Y; terminal X may then choose any suitable combination of audio, video, and data, in the knowledge that terminal Y will be able to receive it.

If both terminals have indicated a capability to receive at higher transfer rates than can be transmitted on the available connection, then terminal A may request one or more additional connections. When each of these is

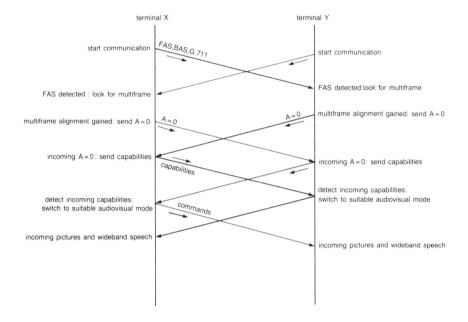

Fig. 7.4 Initialization procedure.

established, FAS and BAS are transmitted in the service channel thereof, and frame alignment is sought; however, the A-bit is not set to zero until multiframe alignment is also found and the channel synchronized with the initial channel. When $A=0$ on additional channels, the terminals are at liberty to expand the communication into those channels by the transmission of appropriate BAS commands.

Initialization between terminals connected to a restricted network (at 56 kbit/s or multiples thereof) is essentially similar to the above. The case of interconnection between a 64 kbit/s and a 56 kbit/s terminal is more complex, since bit 8 is the one which will be lost at the gateway; when the 64 kbit/s terminal has detected that the incoming signal is from a 56 kbit/s terminal, it must switch its service channel to bit 7.

7.4.3 Dynamic variation of the information

Provided that it remains within the capability range indicated by the other terminal, the content of a transmitted signal may be varied at will, using BAS commands. Each command is effective from the next frame-pair; that is, the frame following that in which the error-correction code for the BAS command has been transmitted. Since the frame rate is 100 Hz, a change can be made every 20 ms. Some changes require two BAS commands, the first reducing the rate of one signal type in order to provide capacity for another type, introduced at the second command: for example, to activate data at 14.4 kbit/s in a wideband speech call the sequence {audio 48 kbit/s; data 14.4 kbit/s} must be transmitted.

The CCITT Recommendations provide for transmission modes in which the frame structure is actually switched off: for example, audio or data at 64 kbit/s. In such cases it is necessary to reinstate the frame structure and check the $A = 0$ condition before any mode change can be made.

Prior to disconnection, the communication should be returned to the initial condition of G.711-encoded audio with FAS and BAS; this ensures that, if communication is resumed (for example, after a call transfer or enquiry call), the terminals are ready to do so.

7.5. MULTIPOINT

A multipoint call in a circuit-switched network can be treated as a number of bidirectional point-to-point calls between the terminals and a special bridging unit known as a 'multipoint control unit' (MCU). A basic MCU should transmit to every terminal the mixed audio signals from all other terminals — the mixing is additive and does not involve an increase in bit rate. Pictures of people (as in videoconferencing and videophone) could be treated similarly, spatially adding the pictures from the various locations into a split-screen format. This requires the use of multiple video codecs at the MCU; however a simpler scheme is to switch the video signals such that each participant receives the chosen one of the other pictures (see Chapter 6). Telematic signals can be sent from one terminal to some or all others, but since they are not 'mixed' there has to be contention for the limited transmission capacity available.

The facilities for multipoint working can be greatly enhanced by use of an MLP, but it is recognized that multipoint calls will often be requested from terminals not possessing this capability. The Recommendations in the AV.200 series provide for a basic MCU which can provide audio mixing, video switching, and simple data broadcast between any three or more audiovisual terminals; an MLP is to be defined in AV.270.

It is important that any terminal be capable of connection to an MCU, and communicating with it in exactly the same way as it would with another terminal; it must remain within the capability range declared towards it by the MCU. As an example, consider the case shown in Fig. 7.5.

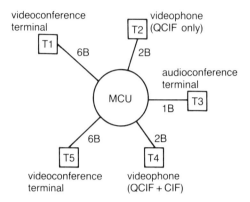

Fig. 7.5 Multipoint connection of dissimilar terminals.

Five terminals are connected to an MCU in star configuration. Table 7.5 shows the terminal capabilities in the left-hand columns: T1 and T5 are identical videoconference terminals, able to work up to 6B (384 kbit/s) with wideband audio, full-CIF video and data within an MLP; T3 is an audiographic conference terminal; T2 and T4 are 2B videophones, T2 being limited to quarter-CIF video format.

Table 7.5 Working mode for five terminals in a multipoint call

Terminal	Capabilities				Working mode			
	audio	video	transfer rate	data	audio	video	transfer rate	data
T1	7 kHz	CIF	6B	6.4 kbit/s	7 kHz	CIF	2B	6.4 kbit/s
T2	16 kbit/s	QCIF	2B	—	16 kbit/s	—	1B	—
T3	7 kHz	—	1B	6.4 kbit/s	7 kHz	—	1B	6.4 kbit/s
T4	7 kHz	CIF	2B	—	7 kHz	CIF	2B	—
T5	7 kHz	CIF	6B	6.4 kbit/s	7 kHz	CIF	2B	6.4 kbit/s

A choice must be made as to the level of service to be provided from the MCU to each terminal. The only common capability between these terminals would be narrowband speech, all other terminals coming down to the level of T2. Clearly this would be an unsatisfactory situation. However, since all the audio signals must be decoded for mixing (a linear operation), it is possible for the wideband terminals to retain the better speech quality, without excluding T2 from the conference.

If T1 and T5 are permitted to work at 6B transfer rate, then only they can use the video facility; however, by lowering the maximum to 2B, T4 can be brought into the video communication. T2 has no full-CIF capability, so unless the other video terminals are willing to have their pictures also reduced to quarter-CIF, T2 must communicate in audio only, requiring therefore only 1B channel. This situation is represented by the right-hand columns in Table 7.5, which show the capabilities declared by the MCU toward each terminal. The values are chosen to prevent T2 from transmitting video and T1,T5 from utilizing more than 2B.

The three teleconference terminals are able to utilize their MLP channel, though this causes a further reduction of 6.4 kbit/s in the video rate. A special code needs to be transmitted to T4 to match this lower video rate.

7.6. CALL CONTROL

7.6.1 Connection

Before audiovisual signals can flow to and from the terminals, suitable channels must be established; we will here focus on the basic ISDN (methods of doing this for broadband (ATM) networks are still the subject of intensive study at present).

The essentials of ISDN call control are set out in [5]. Provision is made for declaration by a calling terminal as to the type of bearer service wanted ('bearer capability', BC) and some wishes as to the type of terminal which should respond at the called ISDN address ('lower-layer capability', LLC and 'higher-layer capability', HLC). For a single-medium terminal this is a simple matter: there is no point in a facsimile machine, for example, addressing anything other than another facsimile, and then only if the two are compatible. For multimedia terminals, audiovisual in particular, there is a wide range of possibilities, as we have seen, and the choices for declaration and response require deeper consideration.

We take the case of conversational applications, though similar conclusions can be reached for other classes such as database access. The most common case may well be that in which a user A simply wishes to address another person B, without knowing precisely what facilities are available to B. In plain language, the declaration from A might mean the following: 'Please provide an unrestricted digital connection (BC), and I would like a response from any terminal which has audiovisual capability of some kind (LLC, HLC); a plain telephone will do if there is nothing better available.' Terminal B must be able to respond to such an incoming request, if that is the wish of its owner. However, it is conceivable that the latter may not wish a highly specialized terminal to become occupied by general calls, in which case that terminal would be programmed to respond only to the exact HLC of interest; probably there would be a general-purpose terminal attached to the same ISDN bus, and it would be this which rings when A calls.

The converse situation may arise, in which it is user A who wishes to specify more closely what kind of terminal he would prefer to respond, setting HLC accordingly (LLC is likely to be more general). For example, he may not want his specialized terminal to be connected through to a plain telephone (answering machine?) if B's specialized terminal is engaged.

If the wishes of the two parties, as expressed by the settings in their terminals, are actually in conflict, there is little that the telecommunications system can do but inform them, as best it can: 'There is no terminal free at the called address which responds to the request you have made'; or 'You requested connection to a Type X device, but in fact a Type Y has responded.' The limited system responsibility undertaken within the standards provision is that at least the various reasonable wishes of users be expressible in the BC, LLC, HLC codes, and adequate capability be included to inform the users as to what is going on. In fact these features must also be extended towards the service provider, who may wish to control and/or be informed for purposes of service-dependent registration or charging.

Adequate code provision to cover these requirements is being made in the 1989-92 study period of CCITT.

7.6.2 Multiple channels

It was seen in section 7.3 that a connection may involve a number of parallel channels established separately. Further call-control codes must therefore be furnished to cover this situation, with added complication as to the choices for request and response.

7.6.3 Transfer

Clearly there must be provision for transfer of a call from one terminal to another at the same address (see also section 7.4.3); the signalling for this is local, so any response restriction involved in the original call set-up can now be overridden.

7.6.4 Multipoint calls

The simplest form of call set-up for a multipoint call is that in which all parties dial into a single MCU, the latter having auto-answer on all ports: no new standards are needed for this. The success of multipoint services will, however, depend greatly on how readily and easily users can access them (as witness the hitherto very low usage of telephone three-party and conference calls). Standards will need to be enhanced to provide the means for:

- changing a point-to-point call into three-party;
- dialling out from or through an MCU;
- linking MCUs together.

7.7. SERVICES AND SYSTEMS

The signal infrastructure of sections 7.2 to 7.4 provides for a wide range of possible communications between terminals, but says nothing about what the preferred modes of communication should be when the terminals are intended for the same service. Recommendations in the AV.100 series describe several audiovisual services largely from the user and service-provider points of view, while the AV.300 series covers the technical systems needed to realize those services as described. Videoconferencing and videophone in particular are covered by Recommendation H.320, which specifies use of the AV.200 infrastructure and identifies the specific transmission modes to be adopted between any two terminals. The stages of progression from a single to a multiple-connection call are also described.

Further standards will eventually be needed for terminals, covering the lighting and acoustic conditions and other human-interface requirements, and these will not be the same for videoconferencing and videophone. It is planned to introduce separate recommendations for these, once there is appropriate experience in the marketplace to call upon.

REFERENCES

1. Recommendations of the CCITT: those relevant to audiovisual services appear in the F.700-series (services), G.700-series (audio coding), H-series (system/infrastructure, video coding), and T-series (telematics); ITU, Geneva (various dates).

2. ISO/IEC JTC1/SC2-WG8, Draft 10918-1: 'Digital compression and coding of continuous-tone still images' (May 1991).

3. CCIR Recommendation 721: 'Transmission of component-coded digital television signals for contribution-quality applications at bit rates near 140 Mbit/s', ITU, Geneva (1990).

4. CCIR Recommendation 723: 'Transmission of component-coded digital television signals for contribution-quality applications at the third hierarchical level of CCITT Recommendation G.702', ITU, Geneva (1990).

5. Griffiths J M, Ed.: 'ISDN Explained', Wiley (1990).

8

PRACTICAL LOW BIT-RATE CODECS

M W Whybray and M D Carr

8.1. INTRODUCTION

Long-distance transmission of uncompressed moving-picture television can be extremely expensive: broadcast-quality television requires in excess of 100 Mbit/s when transmitted in digital form (see Chapter 2). In the early 1980s, the CCITT produced Recommendations H.120 and H.130 which allowed for transmission of pictures suitable for videoconference use at bit rates of 1.5 to 2 Mbit/s. These Recommendations allowed users to buy compatible systems with confidence, and there followed a growth in videoconferencing during the 1980s; however the transmission cost was still a major barrier to widespread use. Non-standard lower bit-rate codecs then started to appear on the market, offering transmission cost reduction but at the price of loss of compatibility.

To address these problems, telecommunications operators and manufacturers of videoconference codec equipment came together within the framework of a CCITT SGXV/1 Specialists Group[1], to devise a new world standard. This Specialists Group initially concentrated on achieving a standard for worldwide videoconferencing at 384 kbit/s and its multiples up to almost 2 Mbit/s. The target date for this standard was late 1988. The group would have followed this with a standard for codecs at 1 and 2×64 kbit/s for ISDN applications. In the event, it was judged important to have the same algorithm for both bit-rate ranges, so after a delay of one year a single $p \times 64$ kbit/s draft CCITT Recommendation was produced which covered all bit rates from 64 kbit/s up to 2 Mbit/s, and became a full Recommendation by accelerated procedure in December 1990.

Video processing technology to implement this standard is now available which can compress videoconference pictures down to a bit rate of only a few hundred kbit/s, and videotelephone-quality pictures including sound down to only 64 kbit/s (the equivalent of a single telephone voice circuit). To achieve these very high compression ratios complex video coding algorithms are employed which tend to be expensive to implement. However, even with current implementation technology, where the capital cost of equipment is of the order of tens of thousands of pounds, the massive savings in transmission costs frequently justify the outlay and many hundreds of companies have now invested in low bit-rate videoconference systems.

This chapter will consider H.261 in more detail and describe equipment which will enable a range of new audiovisual services to become established. In addition, the application of video coding methods to the specific area of sign language transmission will be discussed, particularly with reference to very low bit-rate systems capable of operating on the existing analogue telephone network.

8.2. THE H.261 RECOMMENDATION

8.2.1 Algorithm development

During the five years of CCITT international collaboration to form the new standard H.261, many alternative picture-coding algorithms were explored. Algorithms based on differential pulse code modulation (DPCM), vector quantization, hierarchical techniques, discrete cosine transform (DCT) and hybrid DPCM/DCT techniques were all evaluated. A description of these techniques can be found elsewhere ([2]). All of these picture-coding algorithms are quite complex and it would have been a mammoth task to build real-time hardware for all of them to allow optimization and comparison of the various algorithm performances. For this reason, most of the active parties in the collaboration built up flexible computer-based simulation systems. A video coding simulation system consists of a real-time digital storage medium in association with a mainframe computer and typically includes sufficient storage for a 30 s real-time digitally sampled video sequence. The storage, usually random access memory (RAM) or high-speed parallel transfer disks (PTDs), is first used to capture a moving sequence from a digitized video source. This data is then transferred at a relatively low speed to the mainframe computer. Various computer programs designed to simulate the desired picture-coding algorithm are then executed on the mainframe computer and the processed data subsequently returned to the storage medium. This whole process typically takes several hours for each video sequence. However, after this process it is possible to view the resultant effect of the chosen compression algorithm by replaying the processed sequence at full speed. If desired, radical changes can be made to the algorithm and the new results viewed a few hours later. Many companies throughout the world undertook this work and hundreds of results were compared at the CCITT Specialists Group meetings. After much debate, a coding scheme based on the hybrid DPCM/DCT with motion compensation was selected. The following sections describe the picture format which is an integral part of the coding scheme, and the details of the compression algorithm itself.

8.2.2 Picture format

A major incompatibility which needed to be faced when developing the H.261 standard was that there existed two different line and picture rate standards: 525 lines per picture at 30[1] pictures per second (525/30) is in common use in Japan and North America whilst 625 lines per picture at 25 pictures per second (625/25) is normally used in Europe. To resolve this potential incompatibility, a common intermediate format (CIF) was proposed for H.261 by the CCITT Specialists Group. Both 625/25 and 525/30 codecs include pre- and post-processing modules which convert to and from CIF. CIF is based on 288 non-interlaced lines per picture at 30 pictures per second. Since there are 288 active lines per field (576 active lines per picture) in standard 625 line television, 625/25 codecs have in principle only to perform a picture-rate conversion to meet the 30 Hz picture rate requirement. 525/30 codecs already have the correct picture frequency but instead have to convert the number of active lines from 240. Picture impairments due to this pre- and post-processing are usually negligible when compared with those introduced by the compression coding. In the horizontal direction, CIF is formed by sampling the picture at 6.75 MHz which results in 352 samples (picture cells or 'pixels') per active line. The two chrominance or colour-difference signals in CIF are sampled at half the luminance resolution in both directions, giving a chrominance resolution of 176 pixels horizontally by 144 vertically.

The use of CIF not only solved the compatibility problem but also provided a good picture quality for use in videoconferencing. Broadcast-

---
[1] Actually, the precise picture rate is 30000/1001, or approximately 29.97 pictures/s.

quality pictures contain two sequential fields of 288 active lines; these are interlaced, giving rise to an effect known as 'interline flicker' as well as a significant reduction in the perceived resolution of the displayed image. In fact, subjective studies show that 576 active lines of interlaced video produce the same perceived performance as approximately 400 non-interlaced lines. CIF is therefore capable of reproducing a vertical resolution which is perceived as being only 30% lower than the best possible 625-line studio quality performance. The analogue bandwidth of the luminance is equivalent to 3.3 MHz, about 33% down on full studio-quality analogue television.

CIF is an appropriate choice for many applications, including video-conferencing. However, for some applications (e.g. face-to-face videophone), a lower resolution would suffice. For this reason a second picture format was included in H.261, having 176 horizontal luminance samples per line and 144 lines — one half the resolution of CIF in two dimensions, with corresponding reductions for chrominance. This format is termed quarter CIF (QCIF).

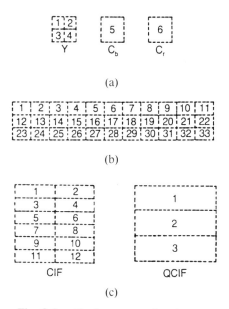

Fig. 8.1 Block structure of an image.
(a) macro-block, consisting of four luminance and two chrominance blocks
(b) group of blocks, consisting of 33 macro-blocks
(c) whole image, consisting of 12 or 3 groups of blocks.

The images are divided into blocks for subsequent processing. The smallest block size is an 8×8 pixel block, but a group of four such luminance blocks, and the two corresponding chrominance blocks that cover the same area at half the luminance resolution, are collectively called a macro-block (see Fig. 8.1(a). Further, 33 macro-blocks, grouped and numbered as shown in Fig. 8.1(b), are known as a group of blocks (GOB), and these in turn are used to build up a full CIF or QCIF picture (Fig. 8.1(c)).

The basic picture rate of CIF and QCIF is 30 picture/s, but a video codec is not constrained to encode every picture. Particularly at the lower bit rates it is usual to omit one or more pictures between each coded one, as a means of helping to reduce the number of bits generated.

8.2.3 The H.261 coding algorithm

A block diagram of the H.261 encoding algorithm is shown in Fig. 8.2. Having obtained an input picture in CIF or QCIF format, a predicted image is then subtracted from the input picture and the resultant difference picture (point A in Fig. 8.2) passed to the discrete cosine transform (DCT) unit. The prediction image is derived from the previously coded image in a manner to be described later. Taking the difference picture substantially reduces the amount of information that needs to be transmitted, since most scenes (especially those of videoconferences) usually contain only fairly small regions of change and all other information in the picture remains approximately constant. After differencing, the picture is divided into 8×8 blocks and then subjected to the DCT process. The transfer function for the DCT is given by:

$$F(u,v) = 1/4 \sum_{x=0}^{7} \sum_{y=0}^{7} f(x,y) \cos[(2x+1)u/16] \cos[(2y+1)v/16]$$

with $u,v = 0, 1, 2, \ldots 7$
where x,y = spatial co-ordinates in the pixel domain
u,v = co-ordinates in the transform domain

A formal description of the DCT process can be found elsewhere [3]. In brief, the DCT produces a series of coefficient values which relate to the spatial frequency content of each 8×8 block. Before transformation, each block is made up of 64 eight-bit values which represent the brightness of each point within the 8×8 space. After transformation, 64 coefficient values result which represent the magnitudes of the various spatial frequency components present at the input to the transform. Each coefficient is then processed in

138 PRACTICAL LOW BIT-RATE CODECS

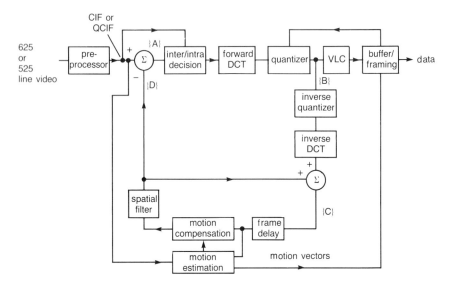

Fig. 8.2   CCITT H.261 encoding loop.

a sequence determined by a zig-zag scanning path shown in Fig. 8.3. The first coefficient value (known as the DC coefficient) represents the mean value (grey level) of the 8 × 8 input block. The second value represents the magnitude of the lowest frequency in the horizontal direction, the third value gives the magnitude of the lowest frequency in the vertical direction, and so on up to the 64th value which gives the magnitude of the very highest frequencies in both the horizontal and vertical planes.

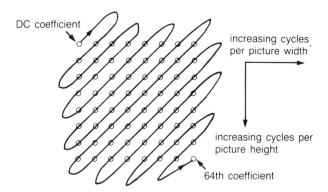

Fig. 8.3   Coefficient scanning sequence for each 8 × 8 block.

The DCT process in itself does not achieve any image compression: 64 values at the input result in 64 values at the output. The compression comes from further processing which takes advantage of redundancy in image statistics. Pictures, and therefore also differences between pictures, tend to be made up of regions of similar values. Most 8×8 blocks can be represented by only a few transform coefficient values: usually only the DC coefficient and a few low-frequency coefficients have significant magnitude. The subsequent quantization process sets all small values to zero and quantizes all non-zero transformed values to a set of nearest preferred magnitudes ready for transmission (point B in Fig. 8.2). Advantage is taken of the fact that for most of the time only a few coefficient values need to be transmitted for each block. Even in the case where a few tens of coefficients need to be transmitted, these values most often occur consecutively within the coefficient sequence and so by using a form of run-length encoding, which exploits the statistics of the occurrence of non-zero coefficient values, relatively efficient transmission is ensured.

Further gain is obtained by the use of variable-length coding [4] which exploits residual redundancy in the coefficient magnitude statistics. The probability distributions of coefficient values have a large peak at zero, and decay away rapidly for progressively larger positive or negative values. Short codewords are therefore used for the commonly occuring values near zero, and longer ones for the relatively rare occurrences of the larger coefficient values. The variable length codewords, plus other control information, are multiplexed together and placed in a data buffer for transmission.

The quantizer is split into a forward quantizer, which maps each input coefficient into one of a finite set of notional levels used to look up the appropriate variable length code in a table, and an inverse quantizer which converts these notional levels back into real (but quantized) numerical values again for subsequent processing.

These quantized transform coefficients are subsequently passed through the inverse transform process and then added to the previously coded image. This results in 8×8 blocks of image data (point C in Fig. 8.2) which are very similar to the video data at the input, though not exactly the same owing to the quantization process. The error appears as noise distributed throughout the block, usually with some spatial frequency structure evident owing to the nature of the transform. However, if the coding parameters are chosen carefully, an acceptable image results. This data is then stored for approximately one picture period ready to be used in the prediction process for the next coded picture.

The motion compensation unit serves to minimize the inter-frame differences which have to be coded. Objects which move simply by translation between sequential images (i.e no rotation, zooming, or deformation) can be most efficiently coded by remapping the object's position from the previously stored image. For example, a person's hand located at a certain point in the previous image can simply be moved to the new position in the current image. Only the displacement vector need be transmitted, resulting in significantly less information than that required to reconstruct the hand itself completely in the new position. Unfortunately, in many situations motion is represented not only by simple translation but also some distortion of the image content by rotation, object deformation etc, so merely moving the previously coded image is rarely more than approximately correct. However, motion compensation does significantly reduce the magnitude of the picture difference signal and thus reduces the bit count. In practice the motion estimation process works by taking a macro-block (a 16×16 luminance pixel area) from the current input video image and searching in the previous coded image over a region of up to ± 15 samples in the horizontal and vertical directions to find a best match. In Fig. 8.2, the motion compensation unit is effectively a variable-length store, whose length is dynamically adjusted so that the best matched position is used for the subtraction (at point D) from the input video image and the values for the horizontal and vertical translations are included as additional data in the output data buffer. Motion is usually only estimated from luminance information, but compensation is applied both to luminance, and, at half the pixel displacement, to the two chrominance components of the macro-block as well.

Since the motion-compensation process is only approximate, the higher-frequency components in particular may be bad matches to, and hence predictions of, the new image data to be coded. The compensation unit is therefore followed by a spatial filter which low-pass filters or blurs the prediction image data if required, thereby removing the poor prediction components. The filter may be switched in or out on a macro-block by macro-block basis, and is typically only switched in if a non-zero motion vector is detected.

The output of the motion compensator and filter processes is the prediction image used by the subtractor at the start of the coding loop. For some macro-blocks where a large change in image content has occurred, even the motion-compensated prediction may be quite poor, in which case the subtraction process may be disabled and the input picture data for that macro-block simply coded directly (intra-frame mode). Since this mode codes a macro-block without danger of corruption from any previously received data that could contain errors, it is also used to update progressively each block in the picture over a period of time, to mop up any errors that may have accumulated due to transmission errors or small differences in arithmetic rounding within the DCT process. H.261 sets specific limits on the DCT error, and on the minimum intra-frame refresh period for macro-blocks.

The different sources of data within the encoder are assembled into an orderly bit stream according to a video multiplex defined within H.261. The multiplex is a hierarchically nested structure, starting with information about the complete frame, then each GOB, then each macro-block, and finally the transform coefficient data. The start of each coded picture is marked by a unique 20-bit codeword that cannot be mimicked by any other video data, and allows the decoder to synchronize to the start of each picture. The following data is then uniquely decodeable to identify which macroblocks within each GOB have been transmitted, and the values of all the associated coefficient values, motion vectors, and so on.

Unfortunately, variable-length codes are very sensitive to errors, and a single bit error can scramble all the data in a GOB before resynchronization is obtained. To counteract this, an additional level of multiplex is added which incorporates a BCH error-correction coding scheme.

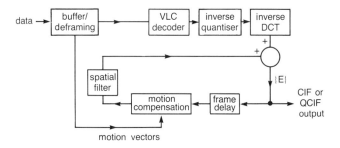

Fig. 8.4   CCITT H.261 decoding loop.

A simplified block diagram of the H.261 decoder is shown in Fig. 8.4. It matches the final part of the coding loop in the encoder, so displacement vectors and coefficient values are input to the decoder and processed in a similar way to the data at point C in the encoder (see Fig. 8.2). If there are no transmission errors then the resulting video data at point E in the decoder will be an exact replica of that at point C.

8.2.4 Video data buffering

The rate of generation of bits by the video encoding process is not constant and depends on the picture statistics at each point in the moving sequence. This variability allows bits to be saved on the 'easy' parts of the image and leaves more bits available for the 'difficult' parts. The number of bits used for each coded picture is also allowed to vary. The network transports data bits at a constant rate and hence video data buffering must be included at the output of the video coding process to provide a smoothing function. Conversely, the decoder utilizes the received information at a non-constant rate when reconstructing the picture sequence and therefore a decoder buffer is also included.

The picture rate on the transmission path has a long-term average value related to CIF. However, if it is measured over a short time interval the picture rate can vary considerably. This variation can be considered as time jitter of the start of each picture. This is illustrated in Fig. 8.5. The buffer size at the encoder and decoder is related to the permitted amount of picture start jitter.

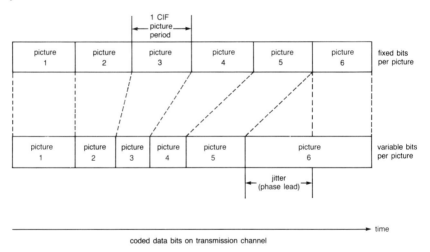

Fig. 8.5 Picture start time jitter.

Earlier video codecs have adopted a fixed encoder buffer size as a means of quantifying the maximum picture start jitter on the transmission path and thus designers have been able to build decoders which can cope with worst-case jitter. In recent years digital signal processors (DSPs) have become available which are powerful enough to be used in the video coding processes.

These coders exhibit a variable processing delay characteristic: if only a few blocks in the picture need processing then the DSP can compute the compressed result relatively quickly but if a large area of picture changes, far more computation time is required. With a DSP-based coder it is therefore usually necessary to place some of the data buffering before the video coding process. Buffering of uncompressed data cannot be easily related to the amount of data that will be created when the picture is subsequently compressed and thus buffer size is no longer meaningful. To allow maximum freedom of codec architecture and the use of DSPs, the Specialists Group chose not to use encoder buffer size as a means of constraining picture start jitter. Instead, the concept of a hypothetical reference decoder buffer has been included. All encoders must be designed to be compatible with this decoder, which is best explained with reference to Fig. 8.6.

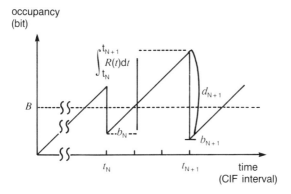

Note: time $(t_{N+1} - t_N)$ is an integer number of CIF picture periods (1/29.97, 2/29.97, 3/29.97, ...).

Fig. 8.6 Hypothetical reference decoder buffer occupancy.

The hypothetical reference decoder buffer is initially empty. It is examined at CIF intervals and if at least one complete coded picture is in the buffer then all the data for the earliest picture is instantaneously removed (e.g. at t_{N+1} in Fig. 8.6). Immediately after removing the above data the buffer occupancy should be less than B. To meet this requirement the number of bits for the $(N+1)$th coded picture d_{N+1} must satisfy;

$$d_{N+1} > b_N + \int_{t_N}^{t_{N+1}} R(t)dt - B$$

where

b_N is buffer occupancy just after the time t_n
t_N is the time at which the Nth coded picture is removed from the hypothetical reference decoder buffer
$R(t)$ is the video bit rate at time t
$B = 4R_{max}/29.97$ where R_{max} is the maximum video bit rate to be used in the connection.

This specification constrains all encoders to restrict the picture start phase lead jitter to four CIF picture periods, worth of channel bits. This prevents decoder buffer overflow in a correctly designed H.261 decoder. Jitter in the opposite direction (phase lag) is not constrained by this specification. Phase lag corresponds to buffer underflow at the decoder which is simply dealt with by making the decoder wait until sufficient bits have arrived to continue decoding.

8.3. PRACTICAL IMPLEMENTATION OF H.261

8.3.1 H.261 flexibility

Significantly, H.261 does not specify a video codec in every detail: there is wide scope for product differentiation without infringement of the basic premise that every codec conforming to H.261 will be able to exchange moving pictures with every other.

The scope for codec differences arises in four main ways.

- A codec may be specified to work at only a restricted range of bit rates, or at all rates from 64 kbit/s up to 1856 kbit/s.

- Within H.261 there are various options for the actual transmitted information. Such options are activated or not according to the mutual capability of the two communicating codecs. Maximum picture resolution capability is one such option — a decoder can demand that only the lowest picture resolution QCIF is used by an encoder. Manufacturers may also design simple codecs which only operate a low picture rate (e.g. 15 or 10 pictures per second). This makes the picture more jerky, but in some applications this is acceptable. A decoder can demand that a certain minimum picture interval is maintained by an encoder.

PRACTICAL IMPLEMENTATION OF H.261   145

- H.261 allows some of the more complex coding techniques to be optional at the encoder: however, decoders must be capable of decoding all of these options. For example, motion compensation is optional at an encoder and mandatory at a decoder. Decoding a motion-compensated signal is relatively simple whilst encoding is quite complex; there can therefore be a real cost saving if a manufacturer chooses not to include this optional feature at the encoder.

- Outside H.261 itself, additional processing steps can be taken which affect the codec's performance parameters: pre-processing in the encoder affects the choice of information to be transmitted in the H.261 format; post-processing in the decoder affects the way such information is used to optimize the display; such processes are proprietary, and naturally affect the cost of the codec. Error correction is another example. Although parity bit calculation at the encoder is mandatory, the more complex error-correction function at a decoder is optional; some manufacturers may choose to sacrifice error performance to cut costs.

This flexibility gives the scope for a wide range of codecs with different capabilities and costs to co-exist, to meet different market requirements. For example, a videoconference codec must produce high-quality pictures of a group of people, and be able to work at a high bit rate, whereas for a 64 kbit/s videophone, lower performance is acceptable but the cost and size become far more important.

The two practical realizations of H.261 codecs designed to meet these two contrasting sets of requirements are described below. One is based on specialized hardware able to operate at full CIF resolution and at any data rate up to 2 Mbit/s. The other is based on general purpose programmable digital signal processors and is a great deal smaller, but can only handle QCIF at up to 112 kbit/s.

8.3.2 Hardware based on H.261 codec

A hardware-based codec is shown in Fig. 8.7. This codec will operate at data rates from 64 kbit/s up to 2 Mbit/s, and both CIF and QCIF picture formats are supported, at the full 30 Hz frame rate. To achieve this high throughput, the codec is specified to be able to process every block in every frame, which requires high-speed digital implementations of all the required functions including the DCT, motion estimation, filtering, and so on. The codec architecture therefore maps directly on to the block diagram of the H.261 coding loop shown in Fig. 8.4. In addition, pre- and post-processing modules in the design ensure that the visibility of coding artifacts is minimized. Error correction at the decoder ensures that good picture quality is displayed even when working over channels with bit error rates as high as 1 in 10^5.

Fig. 8.7 The VC2100 video codec equipment developed by BT.

The codec comprises seven standard circuit cards with room for five expansion cards measuring 35 cm × 23 cm; this includes the audio processing, analogue video processing and channel interfaces. The video codec itself occupies three of these cards. To realize the unit in such a compact form, four application-specific integrated circuits (ASICs) have been developed which help perform motion estimation, motion estimation control, data delay processing and CIF conversion processes. As an example of the level of hardware capability required, the ASIC associated with motion estimation is described in more detail below.

The codec uses motion estimation based on a full search algorithm: every 16 × 16 video input block is exhaustively compared with every possible position in the previous coded frame over a range of ± 15 samples vertically and horizontally (a total of 961 search positions) and the best match found. The computational power required to achieve this function is extremely high. A 16 × 16 video input block occurs every 55 μs. After accessing a new search position, each of the 256 samples in the current image is subtracted from

PRACTICAL IMPLEMENTATION OF H.261   147

the corresponding 256 samples in the previous frame and the mean absolute difference accumulated. 960 similar further operations have to be performed within the 55 μs period. This full-search function is performed by a single ASIC, each having 780 000 transistors in 0.8 μm three-layer metal 'sea of gates' technology and producing a massive 8000 million instructions per second (8000 MIPS). A picture of the chip is shown in Fig. 8.8.

Fig. 8.8   Motion compensation ASIC developed to perform the 'full search' motion compensation process.

Some other parts of the algorithm use commercially available ICs: for example the DCT is implemented using the Thomson/INMOS A121, which can perform forward or inverse DCTs at up to 27 MHz sample rate. Other peripheral operations are performed by programmable logic and gate arrays, and off-the-shelf devices.

8.3.3   Digital signal processor-based codec

For codecs operating at the lower end of the bit range, for example at 64 kbit/s only, and the lower picture resolution of QCIF, then taking account of the fact that the channel limits the number of macro-blocks that need to be processed per second, and by careful use of the optional encoder operations allowed in H.261, the peak processing power comes down to levels that can be handled by general-purpose digital signal processor (DSP) chips.

In this form of implementation, once the input picture has been captured into a frame store, the source coding algorithm is implemented entirely in software running on one or more DSPs. In addition to giving more flexibility at the development stage (since the final details of the algorithm involve no hardware but only software changes), a very compact implementation is achieved that can only be matched in size by a hardware structure having a large investment in VLSI. Using state-of-the-art general-purpose DSPs in conjunction with custom application-specific integrated circuits (ASICs), a low-cost codec on a single printed circuit card is now possible.

The overall architecture of the encoder is shown in Fig. 8.9, and the decoder is essentially similar. The analogue processing stage converts the incoming video format into appropriately filtered and clamped Y (luminance) and C_b and C_r (chrominance) signals. These are digitized by analogue-to-digital converters, and then passed to the frame-store controller module. Here the signals are digitally filtered and subsampled down to the working resolution of QCIF, then written as separate Y, C_r, and C_b areas into part of the RAM. The RAM can hold several pictures at once in different areas, so that whilst one picture is being processed, another can be captured.

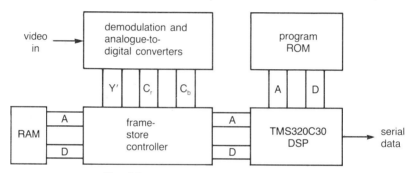

Fig. 8.9 Architecture of a DSP based encoder.

The frame-store controller also allows simultaneous access to the RAM by the DSP, which in this case is a Texas Instruments TMS320C30. This device has an instruction cycle time of 60 ns, during which it can do a simultaneous multiply and add operation, plus various data movements. Even with a QCIF picture resolution, and a maximum working bit rate of 112 kbit/s, it still requires very careful optimization of the use of the processor's time to enable a DSP operating at 17 MIPS to compete with the power of 1700 MIPS mentioned previously for just the motion estimator part of the hardware codec! This can only be done by writing and optimizing the code at the assembly language level, and by reducing the extent to which the optional encoder operations of H.261 are performed. For example, rather than the full search over a range of ±15 pixels in the hardware encoder, the DSP encoder searches over a reduced range, and uses a multistage search whereby a first pass locates the rough direction of motion, and this is refined in a subsequent pass. This compromise slightly reduces the ultimate image quality that can be achieved, as it sometimes fails to find the optimum motion vector, but for its intended application as a videophone codec, this reduced search is satisfactory.

The decoder must of course cope with the full range of operations that an encoder can demand by way of the H.261 bitstream. For example it must be able to motion-compensate over the full range of ± 15 pixels. However, as this is simply a data addressing and moving operation, it is no more difficult than dealing with a reduced compensation range would be. The decoder described does however have a maximum decode picture rate of 29.97/2 (i.e. about 15) Hz, and this, plus the fact that only QCIF pictures can be handled, is agreed with the transmitting codec when the link is established, using the facilities of H.242 as described in Chapter 7. The resulting codec is a single printed circuit card measuring only 18 by 23 cm (Fig. 8.10), with scope for further size reduction, yet fully complying with H.261.

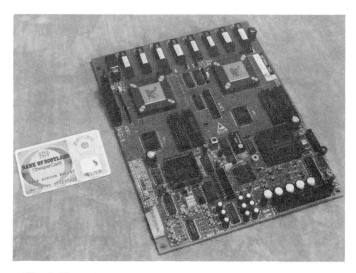

Fig. 8.10 Photograph of a single-card DSP-based video codec.

8.4. PSTN-BASED VIDEOPHONE CODING

Previous sections have discussed algorithms and hardware for videophony at bit rates from 2 Mbit/s down to 64 kbit/s. Below 64 kbit/s it is difficult to maintain a picture quality that would be acceptable to the majority of potential users. However, one specific area where success has been achieved is to provide pictures of adequate quality for deaf people to communicate by sign language. At a bit rate of only 14.4 kbit/s, which can be transmitted over the existing analogue telephone network (the PSTN) using standard modems, two coding algorithms have been developed and shown to provide usable pictures. More background to this application can be found in Chapter 5.

8.4.1 Binary coding algorithm

Early studies by Pearson [5] and Sperling et al. [6] had shown that binary pictures at a low spatial resolution were probably the most suitable format for use with sign language at bit rates around 14.4 kbit/s, as they allowed a frame rate of at least 6 frame/s to be maintained — sufficient to follow most hand motions. Much algorithmic work had been done by computer simulation in laboratories. However, since 1986 the University of Essex and BT have collaborated on the practical implementation of a real-time system that can encode pictures and transmit them over a real network. This work culminated in an experimental system working over the PSTN in February 1989, when two deaf people were able to talk to each other by sign language over the PSTN for what is believed to be the first time anywhere in the world.

The overall structure of the coding algorithm used at the transmitter is shown in Fig. 8.11. Once a new input image is captured, subsequent processing occurs on 8×8 pixel blocks, of which there are 64 in the complete input image. Each input block is combined pixel for pixel with a corresponding one held in a frame-store representing the last coded image, to form a weighted average of the two blocks. This temporally filters the input data to reduce random noise. However, the filter weights are adaptive on a pixel-by-pixel basis and allow large signal variations through unfiltered, thereby avoiding blurring of moving objects.

In parallel with the filtering, the past and present filtered version of the block are compared, to determine if there has been sufficient change in the block data since last time it was coded to warrant updating it by transmission to the decoder. Sixty-four bits, one for each block, are transmitted as a map indicating which blocks are designated as changed. All the blocks so marked pass on to the next stage of the algorithm.

PSTN-BASED VIDEOPHONE CODING 151

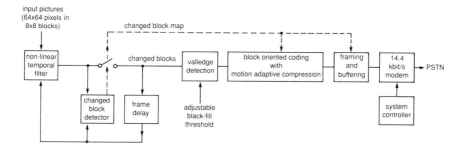

Fig. 8.11 Binary coding algorithm block diagram.

The next process is termed 'valledge detection' [7,8], and is the means whereby the input grey-scale block is reduced to a binary form. An operator is applied to each cluster of nine pixels surrounding and including each pixel in the block, that is sensitive to the presence of any valley-shaped feature in the grey-scale values within each such cluster. The rationale for this is that many significant image features on the human face and hands are not marked by the simple step luminance edge to which conventional edge detectors are sensitive. Instead, many are places where there is a transition between two surfaces of similar luminance, across a narrow region of partial shadow caused by one surface turning tangential to the sight line of the viewer or camera: for example the side of the nose, or the edges of fingers held against the other hand. The resulting small luminance valley is difficult to detect with conventional operators, and also gives double edges, whereas the valledge detector picks these up but also responds to large step edges as well.

The result of applying the above operator is a set of black outlines against a white background. The subjective appearance of such an image is greatly improved by also comparing each input pixel against a grey-scale threshold, and setting all those below the threshold to black, as well as the valley points. This fills in as black any large areas of the image which were dark in the original — for instance a person with dark hair will now appear with black hair instead of simply a black outline filled by white hair. This 'black-fill threshold' is under the control of the user, and is adjusted for optimum picture clarity whilst viewing the self-view picture, which is in binarized form as seen by the other party.

The binarized image block is now compressed using a quad-tree type decomposition. Each 8 × 8 pixel block is examined to see if it is all black, all white, or a mixture of the two (grey), and one of the three following codewords is transmitted to signal the event to the decoder:

Event	Codeword
Grey	0
White	10
Black	11

If the block is grey, it is subdivided into four sub-blocks of 4×4 pixels each, and the same rule is applied to each of these in order. Again, if a block is grey it is further divided into 2×2 blocks, and if necessary into 1×1 blocks, which of course, being single pixels, may be coded using only one bit each to signify black or white. During this process, many blocks are found at the early stages to be all black or white, and no further subdivision of these is necessary. The resulting compression is data-dependent, and relies on the fact that natural images tend to contain large areas of similar luminance and hence result in the early termination of many block subdivisions, with a compression factor of 2 to 1 being typical.

The quad-tree process is wholly reversible, but extra compression can be obtained by allowing some controlled image degradation. This is done by allowing a sub-block to be classed as black or white even if it actually has one or two minority colour pixels within it. This means that the image data is not reconstructed exactly, but allows image quality to be traded against picture update rate, in situations where large amounts of image movement would otherwise result in an unacceptable drop in the frame rate. A consequence of the algorithm is that the amount of data needed to encode each video frame is variable, so a first-in-first-out buffer is used to equalize the bit rate to the fixed line rate. New frames are accepted for processing only when the buffer level falls below a preset minimum; a variable video frame rate is therefore produced. A transmission frame structure, similar in principle to that of H.261, is used to enable the decoder to lock on to the start of each coded frame, and uniquely decode the data.

Moving images from the camera are passed to a video encoder, implemented using a frame store to capture images at a resolution of 64×64 pixels and make them available to a DSP for processing. The architecture is similar to that of Fig. 8.9, but the DSP used is the less powerful TMS32010, as the algorithm is less complex than H.261, and deals with lower-resolution pictures. The complete video codec (encoder and decoder) including power supplies is contained in a unit around $10 \times 25 \times 35$ cm. A conventional data modem is used to provide a full duplex 14.4 kbit/s link over the PSTN.

The general response to this system was very favourable, being found easy to use and providing a means of natural, free-flowing conversation in which humour and emotion were effectively conveyed. However, there was difficulty in obtaining good picture quality in some lighting and background conditions, and it was necessary for users to adapt their signing technique to suit the new medium. Further details of the algorithm may be found in [8]. The algorithm was subsequently improved by adding motion-compensated prediction, as used in H.261, and more sophisticated variable-length coding and adaptation to picture content, which roughly doubled the amount of compression achieved, and hence enabled a higher picture resolution of 96×64 to be used. However, this did not overcome the difficulties inherent in trying to get a good binary representation of the image in the first place, which was still the main problem in using the system. Therefore, attention turned to a grey-scale algorithm, which is described below.

8.4.2 Grey-scale algorithm

Early attempts to provide grey-scale pictures for signing had proved unworkable, as the algorithms used were not as well developed as the binary algorithms, and the target bit rate was the lower one of 9.6 kbit/s, which was the most that could be transmitted over the PSTN at the time. However, the extension of modem rates to 14.4 kbit/s, and the development of motion compensated DPCM/DCT algorithms such as H.261 in particular, gave the necessary ingredients for a workable grey-scale algorithm finally to be realized.

It can be inferred from Sperling's results [5] that a binary picture of resolution $X \times Y$ pixels will have roughly the same intelligibility for signing and finger-spelling as a grey-scale one of resolution $X/2 \times Y/2$, so extrapolating from our binary algorithm work, a grey-scale image of resolution as low as 32×32 pixels would just be adequate. In fact, this was found to be marginally too low, and a final resolution of 48×48 was adopted.

The algorithm developed at BT is based on CCITT Recommendation H.261, but with the resolution reduced to only 48×48 pixels, also using a complex control algorithm to achieve a suitable trade-off of spatial, temporal, and quantization degradations. It is implemented on a reduced specification version of the codec card described in section 8.3.3.

Preliminary subjective tests have been conducted comparing the grey-scale with the binary pictures, using four deaf subjects in pairs doing interactive tasks requiring signing and finger spelling. The time taken to complete tasks was only about 50% greater than for face-to-face with the grey-scale pictures, compared to about 100% longer than face-to-face with the binary ones, indicating a significant advantage for the grey-scale pictures.

Of equal significance were the reactions of the deaf users, who much preferred the natural appearance of the grey-scale pictures, also finding them less difficult to interpret. They found it far easier to recognize the person at the other end, and to observe facial expressions, which are a critical component of sign language.

Interestingly, although users of the binary system had criticized its landscape image format as wasting usable space to the side of the body, and recommended a portrait format, when this was tried with the grey-scale pictures they found it rather constricting as it was too easy to move or sign outside the usable area inadvertently. They finally settled on the square format as the best compromise.

8.5. CONCLUSION

The adoption by the CCITT of a single comprehensive video codec standard for the transmission of moving video at bit rates from 64 kbit/s to 2 Mbit/s assures customers of the satisfaction of interworking between codecs from different manufacturers, and between terminals intended for a range of different purposes. At the same time, manufacturers are free to produce implementations of varying complexity and cost. This article has shown how the scope available in draft Recommendation H.261 can be applied. Also, extension of these ideas to bit rates as low as 14.4 kbit/s has been described; this has found application in enabling deaf people to communicate by sign language over the existing analogue telephone network.

REFERENCES

1. Okubo et al: 'Progress of CCITT Standardisation on 384 kbit/s Video Codec', Globecom 87, Tokyo Japan, pp 36—39 (December 1987).

2. Pearson D (ed): 'Image Processing', McGraw Hill (1991).

3. Clarke R J: 'Transform Coding of Images', Academic Press, London p111 (1985).

4. Huffman D: 'A method for Construction of Minimum Redundancy Codes', Proc. IRE, pp 1058—1101 (September 1952).

5. Pearson D E: 'Visual communication systems for the deaf', IEEE Trans. Commun $\underline{29}$, pp 1986—1992 (1981).

6. Sperling G, Landy M, Cohen Y, and Pavel M: 'Intelligible encoding of ASL image sequences at extremely low information rates', Computer Vision, Graphics and Image Processing, $\underline{31}$, pp 335—391 (1985).

7. Pearson D E and Robinson J: 'Visual communication at very low data rates', Proc. IEEE $\underline{73}$, No 4 (April 1985)

8. Whybray M W and Hanna E: 'A DSP based videophone for the hearing impaired using valledge processed pictures', IEEE Conference on Acoustics, Speech, and Signal Processing, Glasgow, pp 1866—1869 (23—26 May 1989).

9

VARIABLE BIT-RATE VIDEO CODING FOR ASYNCHRONOUS TRANSFER MODE NETWORKS

D G Morrison

9.1. INTRODUCTION

The trend in network switching is towards a multiservices network based on asynchronous transfer mode (ATM) and CCITT has agreed that the broadband ISDN will be based on ATM principles [1]. Such networks can offer a wide range of bit rates without requiring the provision of many separate switching engines and hence are attractive to both network operators and their customers.

For each connection the freedom to use any bit rate within wide upper and lower limits can be exploited by deciding beforehand the required rate and operating at that fixed rate for the duration of the connection. However, for many applications, including video, the possibility to vary the bit rate dynamically during the connection may bring benefits.

9.2. VIDEO CODECS FOR FIXED-RATE NETWORKS

Video coders which employ data compression exploit redundancies in the video signal. As described in Chapter 8 [2], spatial and temporal redundancies are extracted by a CCITT Recommendation H.261 coder [3] as part of the process to compress input video signals requiring 36.5 Mbit/s in uncompressed pulse code modulation form to less than 2 Mbit/s. Because those redundancies depend on the scene content and its motion, the algorithm inherently does not produce data at a predetermined constant rate. For connection to traditional fixed-rate transmission channels, these coders have an output buffer to perform a smoothing and regulating function.

Such buffers, together with the corresponding inverse ones at decoders, introduce delay which must be bounded for interactive applications, less than 100 ms being typical. This period is long enough to smooth variations between the top and bottom parts of the scene but alone would not cope with seconds of little motion followed by seconds of large motion over the whole picture area. To handle this, a fixed-rate coder is forced to adjust its coding parameters dynamically so that the medium-term rate of data generation is within the short-term smoothing capability of the buffer and the long-term average rate exactly matches the transmission channel clock. Commonly, feedback from the state of fill of the buffer controls the step size of a quantization process and in the CCITT H.261 coder this introduces distortion in the form of variable amounts of spatial noise. Additionally, especially when operating at data rates below about 1 Mbit/s, it may be subjectively better to reduce the coded picture rate and introduce motion jerkiness rather than accept the full amount of spatial quantization noise. The coded picture quality therefore is variable for a fixed-rate codec. Coding distortion may be very visible on difficult sections, yet bits may be wasted at other times when the scene has low activity and the coder needs to invoke bit-stuffing procedures to fill the channel.

9.3. ATTRACTION OF VARIABLE BIT-RATE VIDEO CODING

By eliminating the requirement to operate with a fixed bit rate, the control process described above can be omitted, thus permitting the coder to operate with fixed coding parameters and hence constant quality. The use of more bits at times of large picture activity would be offset by other periods of little movement needing many less bits, so that the overall perceived quality would be better than the fixed-rate case but the total number of bits, and hopefully the charge for transporting them, would be about the same. It was therefore logical to suggest connecting a coder to a variable bit-rate network by omitting the buffer. Further advantages would be the elimination of the buffer delay and its cost.

9.4. NETWORK CONSIDERATIONS

Although removal of the buffer and rate-control mechanism from the coder would achieve the video-related goals, there are network factors to consider. Clearly there would be no benefit if the peak video bit rate were used to determine how many sources could be multiplexed on an ATM network, whose links have a fixed total throughput. The network would be under-utilized for much of the time with a direct influence on revenue and hence tariffs. So the concept of statistical multiplexing needs to be invoked. The peak demand times of some video sources will coincide with the troughs of others and in the limiting case of an infinite number of uncorrelated sources the network could be dimensioned using the average bit rate of each. Thus the charge for each user would be related to his mean requirement yet he could have more when needed.

In practice, the number of sources is not infinite, they may not be totally uncorrelated and their average bit rate is not well defined. It is necessary to have statistical models of the sources and use probability theory. This can never predict whether or not the sum of bit rates from all the sources will exceed the network link capacity, but merely indicate the probability. The price to be paid for the ATM flexibility is now apparent. A synchronous network can decide whether or not to accept a request for a new user connection by comparing the rate requested with the unused capacity at that moment. The ATM network carrying variable bit-rate (VBR) traffic cannot accurately predict for any time in the future either the new user's rate or the spare capacity. Thus there is always a finite possibility that the network will be overloaded, the instantaneous total of the sources exceeding the network capacity. The likelihood of this happening increases with the total amount of data which the network is carrying.

Network overload, however brief, causes data to suffer extended delay or to be lost altogether. For some applications, such as file transfers, delay is merely inconvenient and protocols can be used to detect loss and request retransmission. Real-time services such as video and speech may not be able to withstand the resultant delay caused by the overload itself or the retransmission. Data which arrives after the picture has been reconstructed and displayed is no different from lost data. Consequently it is necessary to examine the effect of lost data in a video decoder.

9.5. CONSEQUENCES OF DATA LOSS

Virtually all low bit-rate video-coder schemes utilize temporal redundancy by sending only the changes from one picture scan to the next. These are accumulated in a recursive arrangement at the decoder which is essentially a perfect integrator. Consequently an error introduced at the decoder by data being lost persists for much longer than one scan time and can therefore be very visible, typically for up to 10 s in CCITT H.261. Furthermore, the use of motion compensation can cause these errored areas to expand and move around the screen. Because of the susceptibility of low bit-rate video codecs to transmission errors, many of them incorporate forward error correction which can cope with random or short burst errors as may occur on conventional fixed-rate links. However, ATM networks transport data in packets or cells of 48 bytes, of which four would be taken up by various overheads, leaving 44 for video data. In this case network overload causes data to be lost in multiples of 352 bits. Cell loss is therefore a major consideration when operating a high-compression video codec over an ATM network.

9.6. DEALING WITH CELL LOSS

Several methods to deal with cell loss are under investigation.

9.6.1 Zero cell loss achieved through low network loading

The simplest strategy is to request network operators to keep the network loading sufficiently low that cell loss occurs very infrequently and can be ignored. This may have a very serious impact on the amount of traffic which can be carried and consequently result in high tariffs compared to fixed-rate networks at the same mean rate.

Even assuming that the tariff issue can be solved, the network operator will require information about the characteristics of the variable-rate video data to be carried so that a suitable network loading factor can be determined. For a completely uncontrolled coder which may be given any input scene the possible maximum and minimum bit rates are very different and the shape of the distribution between these limits cannot be predicted. This strategy does not appear to be promising.

9.6.2 Zero cell loss achieved through source characterization and policing functions

An extension of the above strategy is to characterize the bit rates from sources and declare them in advance so that their variability can be better predicted. This allows a better estimation of the total bit rate from all sources so that a higher network loading factor can be used than would be the case if the sources were completely unknown. The difficulty with this approach is that a fully satisfactory way to characterize the sources has not been found. Some success has been reported [4, 5] but scene cuts and document cameras were not included.

Source models with parameters such as mean and peak bit rates require a time interval over which they are measured. The purpose of the characterization is to avoid overflow of the buffers at the network switch nodes, but because these are interconnected and operating on already buffered combinations of sources, it is not a simple matter to relate the characteristic required of any one source to any one buffer. However, it is intuitive that a sensible time interval for defining the source characteristics is of a similar order to that of individual switch buffer storage times.

Bearing in mind that ATM network switch buffers will be at most a few milliseconds, it is apparent that network buffering will not absorb the medium-term variations that a conventional fixed bit-rate coder does in its own buffer of about 100 ms. Because the network buffers are comparatively short, the only mechanism for absorbing the variability is to have sufficient margin of network bit rate available through light loading or statistical multiplexing, preferably the latter.

All this leads to the concept of 'policing functions' to ensure that a source abides by its declared character so that the model on which the statistical multiplexing gain and network loading is based remains valid. The function needs to monitor the use of bits versus time. Several have been suggested and all have a measurement time window or time constant whose value is problematical. If it is very much longer than the network buffer storage times then sources have a great deal of freedom so that it is easily possible for many to peak and overload the network. In this case the policing function is ineffective because it is operating on a macroscopic time frame whereas the network overload is occurring in a microscopic time frame. However, if the policing function time constant is made short then the coder is constrained to keep its variability within short time frames. This is equivalent to saying that its longer-term average rate is bounded and to comply with that it must operate with variable quality.

There seems to be a fundamental incompatibility between a policing function which operates over a short enough timescale to protect the network and the longer-term variability of bit rate necessary to give more freedom to a VBR video codec than possessed by a conventional fixed-rate codec with output buffer. With its equal or better picture quality, the fixed-rate codec may be a better choice because a network with fixed-rate sources can be loaded to 100% capacity and 100% revenue whereas with VBR the mean loading will always be less, possibly substantially so.

9.6.3 Zero cell loss through forward error correction

The most straightforward approach, at least in concept, is to keep the data-loss problem separate from the video coding and apply error correction to the end-to-end link. Classical techniques such as Bose Chaudhuri Hoquenghem (BCH) and Reed-Solomon are not directly applicable to the case of cell loss where hundreds of contiguous bits disappear. The original data from the coder plus the extra forward error-correction bits must be rearranged so that loss of an ATM cell affects a much smaller number of bits in each of many error-corrector blocks. At the decoder, cell loss must be detected and a substitute dummy one generated. The rearranging process at each end introduces delay. Furthermore, there is the possibility that successive cells from one source might be lost and this requires an even wider dispersion of data. Nevertheless, this method may be a good solution where high quality is required but delay is not so important, broadcast television being an example. The design of the error corrector will be influenced by the cell loss probability and, as with the previous strategies, this depends on network loading and source characteristics.

9.6.4 Error concealment

Another approach is to adopt error concealment in the video decoder. Provided the decoder is aware that a cell has been lost it can take defensive action to mitigate the effects, for example by replacing the damaged part of the image from a neighbouring one in space or time. To date, these methods have not been fully satisfactory when the cell loss rate exceeds a relatively low level.

An extension of this approach has been proposed by Wada [6], in which the decoder via a return channel notifies the coder that an area of the picture has been damaged. In this selective recovery method, the coder continues coding but without making backward reference to the affected area, thereby confining the spatial and temporal extent of the corruption. The technique does not add extra delay to the decoded video signal even when the connection has a long propagation delay.

9.6.5 Two-layer coding

An approach currently being studied at BT Laboratories (BTL) is based on work by Ghanbari [7] utilizing two-layer coding schemes which are specifically intended for use on ATM networks supporting at least two levels of priority. Some of the network capacity is allocated to guaranteed cells and the remainder is used for cells which may be discarded if overload occurs. The Orwell protocol [8] devised at BTL is designed to share capacity in this manner and the concept of priority is finding support in CCITT SGXVIII which is setting standards for ATM [9]. Researchers in speech coding for packet networks are also investigating priority schemes [10].

For this two-layer video coding the guaranteed bit rate carries the data for a base-mode picture using high compression. This is known as the base layer. The second layer conveys enhancement data which brings the base-mode pictures up to the required quality. The enhancement layer data is not used recursively and so its loss results only in a momentary localized reversion to the base-mode quality. The technique has been verified by observing pictures derived by computer simulation on a two-layer adaptation of H.261. Even enhancement cell loss rates as frequent as 1 in 10 are not readily visible.

This resilience of a two-layer codec to cell loss prompts contemplation of an increased number of users on the network such that overload and cell loss deliberately occur much more often than would be permitted with single-layer schemes. The higher network loading factor obtained could result in lower charges to the users.

9.7. SIMULATION MODEL

9.7.1 Simulation sequences

Initial investigation of video coding algorithms is normally carried out by computer simulation (see Chapter 8). For reasons of processing time and storage requirements only a small set of short sequences is used. It has been reported [11] that the videotelephone and videoconference simulation sequences used during the H.261 standardization exercise do not exhibit marked temporal variation in their bit rate requirements. Peak-to-mean ratios are about 1.3 compared to around 4.5 for broadcast television [12]. However, actual video meetings do have more variability when participants move around and even slight movement of detailed papers or lighting shadows at document stations has the potential to produce periods of very high bit rates. Some video meetings may contain portions of other material. For example, an advertising agency may wish to show its client a videotape television commercial.

Scene cuts are a good way to introduce large changes in short sequences but the human visual system is temporarily blind to many defects just after the cut, a fact exploited by the fixed-rate codecs.

As a first attempt at constructing a more useful simulation sequence for ATM experiments a new sequence, 'Jack in the Box', has been made. A still from the sequence is shown in Fig. 9.1. The pyramid of rings rocks from side to side and 'Jack' springs out of his box part way through the sequence. The original was captured with a CCD camera (charge coupled devices as sensors) providing red, green and blue signals scanned at the standard European television rates of 625 lines, 50 interlaced fields per second. For coding experiments conversion is first performed to the common intermediate format (CIF) of H.261 which has a luminance signal of 288 visible lines each with 352 picture elements repeating 30 times per second and two colour difference components of half the spatial resolution in each axis and the same temporal rate. These three signals are digitized with 8-bit linear quantization.

Fig. 9.1 A still from the 'Jack in the Box' sequence.

Figure 9.2 is a plot of the bits per picture generated when 'Jack in the Box' is coded by a modified version of H.261 where the rate control has been disabled and the quantizer step size fixed at 1/512 of the theoretical dynamic range of the transform coefficients. The regular motion of the oscillating pyramid throughout the sequence and the large increase caused by the rapid motion when 'Jack' escapes can be clearly seen.

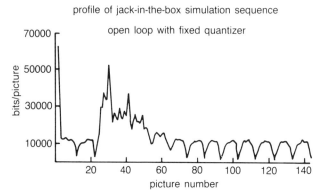

Fig. 9.2 Profile of the 'Jack in the Box' simulation sequence with fixed coding parameter.

9.7.2 Base layer

For the base layer the H.261 scheme was adopted because it represented the state of the art in coding at bit rates between 64 kbit/s and 2 Mbit/s. The algorithm had been well researched with some hundreds of man-years of study in many laboratories contributing to the CCITT standardization activity. Additionally, when new ATM codecs are introduced, it will be desirable that interworking with the then existing population of H.261 codecs be possible. A simplified block diagram of the H.261 codec is given by the solid lines of Fig. 9.3.

The base layer is operated at constant bit rate. This is not essential for picture coding reasons but considerably eases the network dimensioning to ensure that this layer has guaranteed delivery.

9.7.3 Enhancement layer

The enhancement layer carries the difference between the video input and the picture reconstructed by the base layer. This can be derived with a subtractor as shown with dotted lines in Fig. 9.3. Initially there were concerns that this difference data might resemble noise and be difficult to code efficiently, but it was found to contain substantial amounts of correlation and to be amenable to transform coding in the same manner as the base-layer prediction error [13].

Other methods of deriving two or more layers have been suggested [14,15]. The technique used here has the merits of simplicity and minimal changes from H.261. Because the transform is linear the rearrangement to Fig. 9.4 is possible in which the quantization errors of the transform coefficients are extracted. The errors are quantized with a fine quantizer, then zig-zag scanned and variable length coded before being associated with addressing information in a similar manner to H.261. Because of the

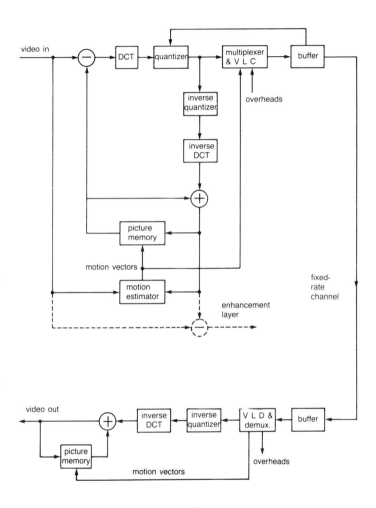

Fig. 9.3 Simplified block diagram of H.261 codec with dotted addition showing extraction of enhancement layer.

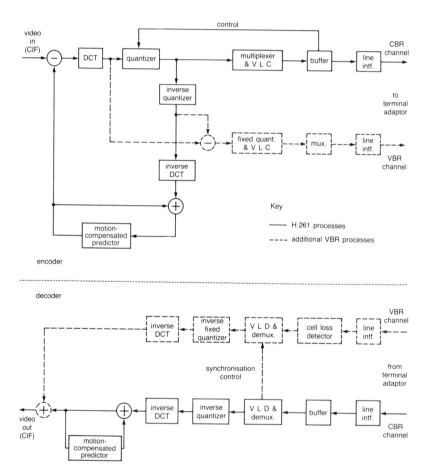

Fig. 9.4 Simplified block diagram of two-layer VBR codec.

possible loss of enhancement cells it is essential that data in each is independent of information in other cells. Consequently the first block address in each cell is coded in absolute form, but any others in that cell are coded differentially to retain as much coding gain as possible. The encoder can ascertain the first address in a cell because it determines the cell boundaries and signals their positions to the cell assembler as described later.

9.8. SIMULATION RESULTS

Results are given here only for the 'Jack in the Box' sequence. These alone are not sufficient to form the basis for valid conclusions, but illustrate the principles involved.

9.8.1 Video coding simulation results

Curve 1 of Fig. 9.5 shows the variation of the luminance signal-to-noise ratio (SNR) corresponding to Fig. 9.2. The 'noise' for the calculation is the pixel by pixel difference between the CIF input to the algorithm and the decoded CIF ouput. The SNR is computed as a mean square error over each picture and expressed as a ratio relative to 255, the peak-to-peak video. As expected from the use of the fixed quantizer, the variation in SNR during the sequence is small. Curve 2 is for the fixed bit-rate H.261 coder using the same mean bit rate of 430 kbit/s and a buffer of 43 kbit corresponding to 100 ms. The dip in SNR of the fixed-rate coder when 'Jack' is sprung is evident.

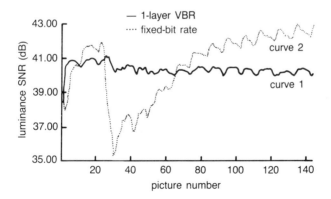

Fig. 9.5 Comparison of luminance SNR for fixed and variable bit rate (one-layer) coding (for the same mean bit rate of 430 kbit/s).

168 VARIABLE BIT-RATE VIDEO CODING

Bit consumption for one configuration of the two-layer coder is shown in Fig. 9.6, curve 2. Curve 1 is a repeat of Fig. 9.2 to aid comparison with the single-layer method. The two-layer coder was allocated 128 kbit/s in the base layer but its total mean bit rate was 1535 kbit/s, more than three times that of the single-layer version.

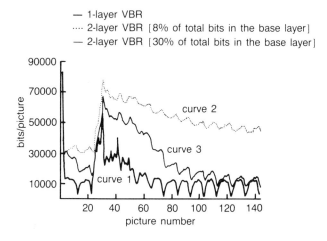

Fig. 9.6 Comparison of a bit requirements for one-layer and two-layer coding.

The coding efficiency of the enhancement layer is not as good as the base layer because the data cannot be used recursively. Therefore the base layer would be allocated as much as possible if coding efficiency were the only goal. Curve 3 of Fig. 9.6 shows the two-layer coder with 256 kbit/s for the base layer, again compared with the single-layer simulation. Total bit consumption is significantly reduced at a mean of 832 kbit/s.

The lower coding efficiency of the enhancement layer is not the sole reason for the higher mean bit rate of the two-layer coder. The data in the enhancement layer cannot be used inside the base-layer coding loop to improve the accuracy of the prediction. The enhancement layer may therefore be continuously mopping up errors left uncoded by the base layer whereas in a single-layer coder bits would be expended just once to code those errors. For example, stationary background may never be exactly coded by the base layer because the amount of movement elsewhere may constrain the quantizer step size to be larger than the background errors. These then have to be coded repeatedly by the enhancement layer on every picture scan.

It has been found possible to reduce this effect significantly by forcing the base layer to use a fine quantizer occasionally on parts of the picture in turn. Although this generates more bits which cause other moving parts to be coded less well by the base layer and consequently an increase from those areas in the enhancement layer, it is a once-off penalty which is more than repaid by the consequent removal of that area from continuous contributions to the enhancement-layer data.

Simulation results with the forced use of a fine quantizer in the base layer are shown in Fig. 9.7. The bit rates are listed in Table 7.1.

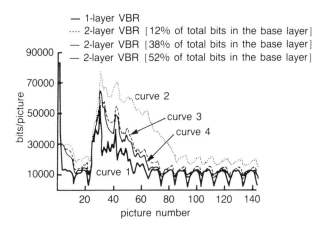

Fig. 9.7 Comparison of bit requirement for one-layer and two-layer coding with forced use of fine quantizer in the base layer.

Table 9.1 Bit rates of curves in Fig. 9.7.

Curve	Base layer (kbit/s)	Peak (kbit/s)	Total mean (kbit/s)
1	0 (single layer)	1685	430
2	128	2341	986
3	256	2001	654
4	320	1935	594

Figure 9.8 shows the corresponding SNRs which are all between 39.6 and 41.6 dB with much reduced temporal variations compared to the fixed rate result in Fig. 9.5.

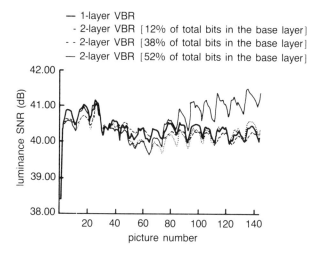

Fig. 9.8 Comparison of luminance SNR corresponding to Fig. 9.7.

9.8.2 Network loading simulation

It is important not to concentrate solely on the coding efficiency as measured in terms of picture quality against bit rate. Although that can be a valid comparison between schemes for synchronous networks, in the case of ATM networks it is possible that the two-layer algorithm could permit more users on the network than a single-layer algorithm even though the average bit rate of each user is higher. This is a direct consequence of its superior resilience to cell loss allowing a higher network loading factor.

Some preliminary simulations have been carried out to test this hypothesis that the use of more bits by the two-layer coder can be offset by the relaxed cell-loss rate requirement of the enhancement layer. The results were obtained from an approximate analysis of the loss probabilities of ATM cells based on the mathematical theory of large deviations. This theory yields an upper bound on the cell-loss probabilities, but other work has shown that the upper bound provides an approximation which is generally accurate. The analysis assumes that the cell stream produced by the video codec is bursty, consisting of periods when cells are emitted at peak rate alternating with periods when no cells are generated. Under this assumption the stream of cells is completely specified by its mean and peak rates. The network is assumed to be bufferless, with the result that cells which cannot be transmitted immediately are lost.

Figures 9.9, 9.10 and 9.11 plot the cell-loss ratio versus the number of simultaneous connections on networks with total available capacities of 10, 30 and 140 Mbit/s respectively. The VBR source in each connection is modelled only by the peak and mean bit rates from the 'Jack in the Box' simulations as described above. Curve 1 is the single-layer coder and curve 2 is the two-layer coder which has 320 kbit/s for the base layer.

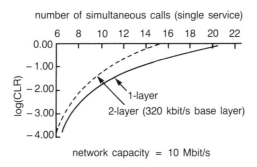

Fig. 9.9 Cell loss ratio versus loading of a 10 Mbit/s network.

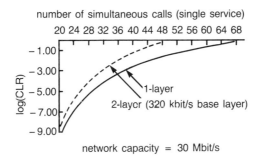

Fig. 9.10 Cell loss ratio versus loading of a 30 Mbit/s network.

Interpretation of the curves to compare the single-layer and two-layer VBR coders is not straightforward because different cell-loss ratios should be used. Subjective tests need to be performed to determine cell-loss ratios which yield equal subjective quality. For the single-layer coder operating at 430 kbit/s mean a cell-loss ratio of 10^{-5} will cause very visible disturbances every 90 s on average and this could be regarded as an upper bound on

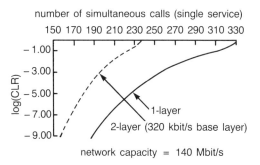

Fig. 9.11 Cell loss ratio versus loading of a 140 Mbit/s network.

the cell-loss ratio of 10^{-6} or better aimed for. The two-layer coder could be operated with a cell-loss ratio of around 10^{-3}.

At a total network bit rate of only 10 Mbit/s the potential for statistical multiplexing gain is small and both techniques offer about six simultaneous connections as shown in Fig. 9.9. The equivalent bit rate of each source is thus very similar to the peak rate of 1685 kbit/s from the single-layer coder. When the network bit rate is 30 Mbit/s (Fig. 9.10), about 28 simultaneous connections can be supported, equating to 1070 kbit/s each. With the above-mentioned cell-loss ratios the two-layer coder can offer 19% more connections than the single-layer coder.

Figure 9.11 indicates that on a 140 Mbit/s link around 200 simultaneous connections can be permitted. Significantly more statistical multiplexing gain is obtained and the equivalent bit rate is reduced to 700 kbit/s. Under the same assumptions of cell loss ratios, the single-layer coder is able to make 10% better use of the network.

Even the most favourable of the above results does not conclusively prove the case for VBR video coding. The single-layer coder with a mean bit rate of 430 kbit/s and the two-layer coder at 594 kbit/s mean should be compared subjectively with a fixed-rate coder operating at 700 kbit/s. All three permit the same number of simultaneous connections and therefore incur similar charges.

9.9. EXPERIMENTAL HARDWARE CODEC

An experimental real-time VBR video codec has been constructed at BTL, to further the VBR studies, though extensive results are not available at the time of going to press.

9.9.1 Experimental video coder

The algorithm implemented is that described previously for the simulation experiments. The electrical design is largely based on dedicated logic functions constructed from programmable devices though some application-specific integrated circuits are incorporated for motion vector search and the transforms [2].

9.9.2 Connection to the network

A VBR codec can be connected to an ATM network in a number of ways depending on whether the codec is a broadband ISDN device (TE1) or not (TE2). The experimental codec can be regarded as a TE2 device and consequently requires a terminal adaptor (TA) to perform the ATM adaptation layer (AAL) function. The design of the TA takes into account the practical considerations of its location in the network and the distance between it and the codec. In the experimental configuration the TA is close to the ATM node.

Because the coder inherently produces two outputs, it was expedient to provide two physical interfaces between the coder and the TA. It would be possible to combine the two data streams and use only one physical interface, but separation of some form would still be needed at the TA to implement separate queues for the base and enhancement cells on to the network.

As shown in Fig. 9.12, the base-layer data is transferred to the TA at constant bit rate across the CBR interface and the enhancement-layer data across the VBR interface. Both are synchronous and use a bipolar line code, the CBR one operating at 2048 kbit/s and the VBR one at 8192 kbit/s. Base-layer data may occupy a predetermined bit rate up to 1856 kbit/s in discrete 64 kbit/s steps as is normal for a H.261 codec. The enhancement data may vary dynamically between zero and a maximum of 6850 kbit/s.

As mentioned earlier, the cell segmentation points are determined by the video coder and these must be indicated to the TA. For this purpose an 8B9B block line code is used, solely between the codec and TA, to provide unique marker codewords which cannot be emulated by video data. Further codewords are utilized in the mechanism whereby the variable-rate enhancement-layer data is conveyed across the 8192 kbit/s interface. Dummy segments are created as necessary, preceded by codewords which indicate to the TA that the data is for filler purposes and should not be launched on to the network.

9.9.3 ATM adaptation layers

As the CBR channel has guaranteed transmission, its ATM adaptation layer functions are segmentation, cell assembly, disassembly and serialization. The adaptation layer function for the VBR channel has the additional task of inserting cell numbers so that the decoder can detect cell loss and take defensive action. If a cell has been lost, data in the next one is discarded until a block address is detected, thus ensuring that no erroneous block data is decoded by concatenation of cells which originally were not contiguous.

In the experimental equipment the cell numbering is actually performed by the video coder and a programmable random cell destroyer is included in the decoder. These features allow codec testing and cell loss experiments to be performed without a network.

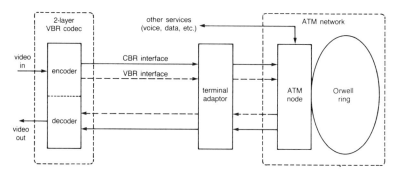

Fig. 9.12 Connection of the codec to the network.

9.9.4 Cell jitter and clock recovery

The base-layer codec retains the rate-smoothing buffers of a fixed-rate codec though their size, and hence delay, can be reduced somewhat. The base-layer receiving buffer at the decoder removes the arrival jitter of the base-layer cells. The H.261 recommendation, although intended for synchronous networks, does not explicitly provide a mechanism for recovering the video clock, so this aspect is no more difficult with ATM. Enhancement-layer data is also buffered at the decoder such that when decoded it can be added in the correct time relationship to the base layer.

9.10. TARIFFS FOR VBR VIDEO ON ATM NETWORKS

To date, the work on ATM networks and video coding for them has concentrated on technical issues. The question of tariffs will arise: no proposals are made here, but it is pointed out that the cost and the quality of a service are likely to be the factors on which the user makes his assessment. The fact that the carrier is transporting the signals over a synchronous or asynchronous network is of secondary interest. The viability of ATM networks and variable bit-rate video on them will depend on the charging basis. In the case of two-layer video coding it will be important that the ratio of charges between the guaranteed and non-guaranteed cells is roughly in line with the coding efficiency of the two layers, otherwise codec designers will optimize the balance of use of the two layers to minimize the total charges and perhaps result in an unsatisfactory mixture from the overall network point of view.

9.11. CONCLUSION

The motivation for variable bit-rate video coding has been outlined and the BTL activity on the two-layer approach described. The topic needs to be considered from an overall system standpoint which includes both the network and codec design. This is in contrast with synchronous networks and fixed-rate codecs where the two can be decoupled. Investigations are still at an early stage as is the work of many others throughout the world. The case for VBR video on ATM networks is not yet conclusively proven; clearly any benefit over alternative (fixed bit-rate) strategies will depend on the tariff approach as well as the technical design.

ACKNOWLEDGEMENTS

The author acknowledges the continuing contributions by Dr M Ghanbari at the University of Essex with the early encouragement by M D Carr of BTL. The coding simulations and the implementation of the hardware codec were a team effort by D O Beaumont, I Parke, A P Heron, S R Gunby, M E Nilsson, N E MacDonald, E G Sinclair and R S Craigie. P R Robinson and I P Thomas were responsible for the Terminal Adaptor and the investigation of network loading versus cell loss rate was by Dr J M Appleton and T R Griffiths.

REFERENCES

1. Draft CCITT Recommendation I.121(Rev): 'Broadband Aspects of ISDN', (January 1990).

2. Whybray M W and Carr M D: 'Practical low bit-rate codecs', Chapter 8 of this book.

3. Draft CCITT Recommendation H.261: 'Video Codec for audiovisual services at px64 kbit/s', to be ratified July 1990.

4. Maglaris B, Anastassiou D, Sen P, Karlsson G and Robbins J: 'Performance models of statistical multiplexing in packet video communications', IEEE T-COM, 36, pp 834—844 (July 1988).

5. Nomura M, Fuji T and Ohta N: 'Basic characteristics of variable rate video coding in ATM environment', IEEE J-SAC, 7, No 5, pp 752—760 (June 1989).

6. Wada M: 'Selective recovery of video packet loss using error concealment', IEEE J-SAC, 7, No 5, pp 807—814 (June 1989).

7. Ghanbari M: 'Two-layer coding of video signals for VBR networks', IEEE J-SAC, 7, No 5, pp 771—781 (June 1989).

8. Falconer R M and Adams J L: 'ORWELL: A protocol for an integrated services local network', Br Telecom Technol J, 3, No 4, pp 27—35 (1985).

9. Draft CCITT Recommendation I.150: 'B-ISDN ATM Aspects' (January 1990).

10. Petr D W, DaSilva L A and Frost V S: 'Priority discarding of speech in integrated packet networks', IEEE J-SAC, 7, No 5, pp 644—656 (June 1989).

11. Ghanbari M and Pearson D E: 'Components of bit rate variation in videoconference signals', IEE Electronics Letters, 25, No 4, pp 285—286 (February 1989).

12. Ghanbari M and Pearson D E: 'Variable bit rate transmission of television signals', IEE Electronics Letters, 24, No 7, pp 392—393 (1988).

13. Ghanbari M: 'An adaptive video codec for ATM networks', Third Packet Video Workshop, Morristown NJ (March 1990).

14. Kishino F, Manabe K, Hayashi Y and Yasuda H: 'Variable bit-rate coding of video signals for ATM networks', IEEE JSAC, 7, No 5, pp 801—806 (June 1989).

15. Karlsson G and Vetterli M: 'Packet Video and its integration into the network architecture', IEEE J-SAC, 7, No 5, pp 739—751 (June 1989)

10

STANDARDIZATION BY ISO/MPEG OF DIGITAL VIDEO CODING FOR STORAGE APPLICATIONS

D G Morrison

10.1. BACKGROUND

Images contain a great deal of information. This is the fundamental problem to be overcome when storing video signals and is essentially the same one as is encountered when transmitting them from one place to another. Conventional television, of the form viewed daily by hundreds of millions of people, requires around 5 MHz of analogue bandwidth or 216 Mbit/s in the digital format recommended for studio use. Comparison with the 15 kHz bandwidth of an audio cassette or the 11.5 Mbit capacity of a 3.5 in 1.44 Mbyte floppy disk as used with modern personal computers gives some insight as to why the introduction of the domestic video recorder lagged so much behind its audio-only counterpart and why pictures on computer screens have been restricted to stills or cartoons with limited motion.

In the late 1950s practical videotape recorders began to appear for routine use in TV studios, the key innovation being rotating magnetic heads to yield the high tape-to-head relative speed needed for video bandwidths while keeping the linear tape speed down to a modest figure. By the early 1980s the technology had progressed to the stage where video cassette recorders (VCRs) could be sold at sufficiently low prices to attract home users in large numbers.

In parallel with developments in videotape recording, much attention was being paid to disk systems [1]. Although these did not offer recording facilities to the end user they held the promise of significantly lower manufacturing cost for pre-recorded material because disks could be pressed in a similar manner to audio LPs. Several competing systems were developed but the most notable and the only long-term survivor was Laservision from Philips employing analogue frequency modulation to vary the spacing of minute pits in a 300 mm diameter disk, rotating at 1500 rev/min and giving an hour of playing time. Although this gave better picture quality than contemporary VCRs, for various reasons it never achieved the same popularity in the consumer market place. However, it found some success in the industrial sector particularly when players were fitted with external control ports to permit the functions, such as random access, necessary for interactive applications such as training programs. A significant by-product was that the technology involved in pressing the disks and in reading the information from them contributed heavily to the development of the audio compact disc.

By the late 1980s the technological feasibility of digital video recording able to reach the demanding quality requirements of television broadcasters had been proven and equipment began to appear. The basic approach was to refine the existing mechanical arrangements and change the electronic processing to handle pulses at 216 Mbit/s instead of 10 MHz analogue frequency-modulated signals. Although apparently simple in theory, the practice demands complex engineering which together with the relatively small market for such machines results in their costing over 100 times as much as home VCRs. The tape costs per hour are in a similar ratio.

Also by the late 1980s two separate strands of technology were becoming available to tackle video recording in a different way. The CD-ROM had been developed from the audio CD to provide low-cost mass data storage. It offers in excess of 600 Mbytes and can be stamped quickly and cheaply just like the audio version. However, even 600 Mbytes is only enough for 22 s of video at 216 Mbit/s and the transfer rate of CD-ROM drives is also limited, typically being 150 kbyte/s. The answer to these two problems of capacity and throughput was motion video compression coding which was being actively researched to facilitate videotelephone and videoconferencing services with relatively low transmission charges [2].

Among the first to bring moving picture compression and CD-ROM storage together was a group at the RCA Laboratories. The digital video interactive (DVI) system originated there and was subsequently taken through to the marketplace with its proprietary algorithm by Intel [3]. As the initial cost of playback equipment was too high for home users, the corporate market sector was the first to be addressed.

10.2. ESTABLISHMENT OF THE MOVING PICTURE CODING EXPERTS GROUP

The above-mentioned technical opportunities, the convergence of computers, communications and entertainment, in addition to the desire to have standards in place before volume sales of incompatible proprietary solutions, prompted the establishment of the Moving Picture coding Experts Group (MPEG) in the spring of 1988. A part of the International Standards Organization, its role was to perform the equivalent for moving pictures to that which ISO JPEG was carrying out for still images [4]. The significance of this activity was quickly realized by major players from the consumer electronics, computer, telecommunications and integrated circuit industry sectors and the number of participants increased rapidly.

MPEG defined three work items to be tackled sequentially. The first was for storage media having a throughput of $1-1.5$ Mbit/s, the second for $1.5-5$ Mbit/s and the third for $5-60$ Mbit/s. The rates for the two later items have since been revised. For a brief period the mandate covered video coding only but it was soon realized that users would expect to have accompanying audio. Since CD-quality stereo audio, with its 16 bits per channel at a 44.1 kHz sampling rate, consumes all of the bit rate addressed by the first work item then audio compression would be needed too and a group was formed for that. Another group was formed to handle the various system issues such as the multiplexing method to interleave the video, audio and any other data.

The target bit rate of the first work item was heavily influenced by the ready availability of CD-ROM as an existing and established data carrier. By coincidence, it was also about the same as the 1.5 Mbit/s or 2 Mbit/s rate of the first level of multiplexing in digital telecommunications networks and thus offered scope for new moving image services on them. Although digital moving video on CD-ROM was probably the dominant perceived application it is important to note that MPEG did not set out to meet just that goal. Instead, it sought a generic standard that would be applicable across a wide range of storage devices and applications. For example Data-DAT, derived from the digital audio tape standard, also has a throughput in the 1.5 Mbit/s region, allows up to two hours per cassette and, unlike CD-ROM, permits easy recording by the end user. Hard disks in personal as well as mainframe computers also have sufficient capacity for worthwhile durations of coded video.

By the time that MPEG was established the CCITT SGXV Specialists Group on coding for visual telephony was within a year of finalizing its technical work towards Recommendation H.261 covering a video codec for audiovisual services in the range $64-1920$ kbit/s [2,5]. Both computer

180   DIGITAL VIDEO CODING

simulations and early experimental hardware codecs were demonstrating the quality which could be achieved at 1 Mbit/s from that algorithm. While appreciating the large amount of good work which the contributors to CCITT had carried out, MPEG recognized that its requirements were not exactly the same. Consequently MPEG decided to seek an algorithm more fully aligned to its own requirements.

10.3. COMPETITIVE PHASE

It was common practice in the ISO and other standardization bodies to produce a standard based upon an existing solution, often one which had achieved *de facto* status through success in the marketplace. Mindful of the format war in home VCRs and the failure of quadraphonic sound, partly because of many competing and incompatible systems, MPEG, in the same manner as CCITT was successfully doing with its work towards H.261, sought to be the forum for the joint development of a standard by all interested parties. It decided to begin with a competitive phase to bring existing ideas to the table and to stimulate the creation of new ones.

10.3.1   Assessment of proposals

Given MPEG's desire to produce a generic standard and the wide range of participating organizations and their interests, it was clear that establishing a definitive set of requirements through discussion alone was not likely to be accomplished easily or quickly. A quantitative procedure was therefore devised. First, participants listed the possible applications and these lists were then collated to produce an overall applications list which would be more comprehensive. Examples included electronic newspapers, surrogate travel, catalogue shopping, games, multimedia art, education, and video plus audio databases. From this list a second one of technical implications was constructed containing items such as forward playback, reverse playback, fast forward, fast reverse, random access, still mode, compatibility with computer graphics, real-time encoding, performance after multiple encoding and decoding passes, editing, error robustness, hardware complexity, cost and so on.

Ballot papers were then distributed containing both the application list and an implication-versus-application matrix. Respondents rated the applications using a scale of 0 to 5, covering 'not important at all' to 'most important application' respectively. Each box of the matrix was marked 0, 1 or 2 indicating 'not important or not related', 'a desirable but optional implication' and 'an implication of very high importance for this application'.

By multiplication and summation over all applications a set of numbers was obtained for each technical implication and these, after averaging over all the respondents, were to be used as weights to apply to tested features of proposed algorithms to derive a final figure of merit for each one.

In the event, the number of technical implications was 28 which would have consumed much effort by the individual proposers and the assessors of the 15 proposals. Also, some of them would have been difficult to measure accurately. Therefore only a subset containing the higher weighted implications was used as follows; where the scores for all these parameters were normalized to the range 0 to 10.

Figure of merit =	quality of normal playback	× 108 +
	quality of reverse playback	× 50 +
	quality of high resolution still	× 76 +
	random access	× 118 +
	CCITT H.261 compatibility	× 26 +
	JPEG compatibility	× 47 +
	VLSI implementability	× 103

10.3.2 Scoring of parameters

The quality scores were obtained by subjective testing. As explained elsewhere [2] it can be exceedingly expensive and time-consuming to implement coding algorithms with real-time hardware and it is common practice to employ computer simulations. MPEG selected three short sequences totalling 25 s. Proposers were required to submit these sequences as coded and decoded by their algorithms, together with files of the compressed versions and executable code for performing the decoding process. The latter two items were for verification if that would prove necessary later.

The random access score (r) was derived mathematically from the longest decoding time (d) to any individual picture in the sequences using the formula:

$$r = 10 - 5d$$

Delays longer than 2 s would receive a zero score.

The compatibility scores were difficult to apportion. In the event, those algorithms which employed a similar structure to H.261 and JPEG with an 8×8 discrete cosine transform (DCT) were allowed the full 10 points and others received 0.

The VLSI evaluation concerned only decoders as these were considered to be much more numerous than encoders. Derivation of the procedure was not straightforward as it was realized that the only way to obtain accurate measures would be to design and produce real devices. As this was not a realistic burden to place on proposers it was agreed that they should provide outline designs including the overall architecture, device partitioning, amount of memory, etc. The total cost figure was then computed by means of the formula:

Cost = cost of application specific ICs (ASICs)	×0.27+
number of Mbits of dynamic memories (DRAM)	×9.96+
number of VLSI chips including DRAMs	×3.54+
total number of pins of all chips	×0.0298

The ASICs were designed using a 'MPEG reference technology' so that all proposers were on an equal footing. The basic parameters of this reference were formed by averaging the confidential answers to a questionnaire sent to VLSI companies.

Although the complete proposal evaluation methodology was agreed by the MPEG participants there were some doubts as to its ability to produce true and meaningful results. Fortunately, the purpose was not to pick a clear winner but rather, through the spirit of competition, to encourage proposers to produce excellent ideas which could then be pooled in the following collaborative phase.

10.4. COLLABORATIVE PHASE

As mentioned earlier the CCITT SGXV Specialists Group on coding for visual videotelephony and the ISO/JPEG committee had both been very active immediately preceding the establishment of MPEG and were showing some impressive results using the DCT. It was therefore not surprising that all but one of the proposals to MPEG used the DCT, the majority being based on CCITT H.261 and some on JPEG, both with some extra features or modifications. The two main areas where new techniques had been incorporated were in the provision for random access and in modifying the predictor to yield higher compression for the same visual quality. However, there were several means of achieving these and it was not clear from the results of the competitive phase which individual methods were best and whether any further benefit could be obtained by applying them simultaneously.

To resolve these issues a simulation model was defined. The first version was not unique, having several options to encompass all the techniques which had shown promise in the subjective tests. By performing carefully controlled simulation experiments the various options were compared and the simulation model gradually refined. Although it was tempting to make decisions just from the results of these simulations, it was recognized that some options might have a large enough impact on the VLSI implementability to outweigh the increase in picture quality and so the VLSI group continued to contribute.

10.5. THE MPEG VIDEO CODING ALGORITHM

The following section describes the salient points of the algorithm as at October 1991. The algorithm had been frozen since September 1990 except for a few errors which came to light. Documentation [6] was in the final stages of preparation, to be released for comment outside MPEG once verification by real-time hardware of the complete MPEG system comprising demultiplexing, video decoding, audio decoding and delay synchronization of video and audio had been demonstrated. Such verification was deemed essential as MPEG did not wish to issue a draft standard which might be flawed. This stance was partly based on the experience of the CCITT where hardware verification trials of the draft Recommendation H.261 had revealed an inadequacy which had not been uncovered by the simulations.

It is important to note that although the algorithm was developed with the 1.5 Mbit/s total bit rate in mind, it is not restricted to that rate. The following description will use as examples the picture resolutions which are well matched to that bit rate but MPEG is applicable over a wide range of picture formats. This is a source of many strengths of the algorithm but may also appear as a weakness in that there is no unique form of MPEG which an equipment manufacturer can target.

Conceptual block diagrams for the encoding and decoding processes are given in Figs. 10.1 and 10.2. These do not necessarily correspond to real implementations. Comparison with equivalent diagrams for H.261 reveals great similiarity of structure. In fact, the MPEG algorithm has much in common with H.261 and may be considered as a logical superset of it though the two are not compatible at the bitstream level. It is not possible to construct a valid H.261 bitstream which is also a valid MPEG one and vice versa. However, many basic parameters such as the DCT block size, the DCT specification, some code tables and so on are identical. These were retained not just to avoid reinventing a wheel but also because this approach offered a higher probability that VLSI devices could be designed which would be switchable or configurable to implement the MPEG, H.261 and JPEG algorithms.

184 DIGITAL VIDEO CODING

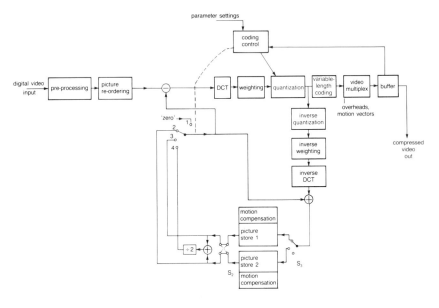

Fig. 10.1 MPEG video encoder.

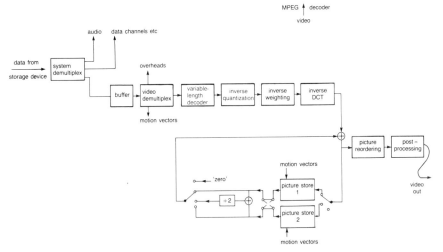

Fig. 10.2 MPEG video decoder.

10.5.1 Pre-processing and post-processing

As mentioned earlier, conventional television requires 216 Mbit/s when digitally coded with pulse code modulation (PCM) in accordance with the

4:2:2 format of CCIR 601 [7]. This figure arises from sampling the luminance component at 13.5 MHz, the two colour-difference signals at 6.75 MHz each, and digitizing with eight bits. It is possible to reduce this gross amount to a net value of 162 Mbit/s by omitting the blanking intervals which contain no picture information but allow time for the scanning spot in the display cathode ray tube to return from the right- to the left-hand side and from the bottom to the top. Further reduction by more than 100:1 is required.

As was well known from the CCITT studies towards Recommendation H.261 the best overall quality at around 1 Mbit/s is not obtained by attempting to encode the TV studio-quality pictures directly. All the proposals to MPEG followed an analagous route to H.261 by first reducing the spatial and temporal resolutions. Because the CCITT algorithm is intended for universal use in audiovisual communications services it addressed the practical difficulty that there is not just one television-scanning format in use throughout the world, but two — 625 and 525 lines with respectively 50 and 60 interlaced fields per second. As mentioned above the allowance for display flyback times means that only 575 or 485 lines carry visible picture information. To permit connections to be established without users, having to worry about the need for conversion facilities, H.261 internally employs the common intermediate format (CIF) and its simply related, downsized version quarter CIF as the only permitted formats which the coding kernel acts upon [2]. CIF having 288 lines repeated 30 times per second not only contains less information to be coded but is also a compromise between 625/50 and 525/60. The conversions from and to the local television format, which can be performed in a variety of ways and are purposely not defined in H.261, can introduce some artefacts, especially motion blur or judder when changing the field rate, but the level of these effects is low with typical videoconferencing and videophone images and very often masked by other distortions from the compression coding.

MPEG, however, was aiming at a somewhat higher picture quality and a much wider range of input picture material. Another consideration was that some applications would be based on computer display formats of which there are several and they differ from 525- and 625-line television. In accordance with the desire for a generic standard the CIF approach was therefore not followed. Users have the freedom to adopt the format best suited to their needs, the only common factor being progressive scanning. Accordingly the formats most likely to be used for television images are 288 lines repeated 25 times per second and 240/30, both having 352 picture elements on each line. These numbers are for the luminance component, the chrominance ones having half the spatial resolution in each axis, and are simply related to the 625- and 525-line formats to aid the conversions. With 8-bit PCM coding the pictures presented to the compression encoder proper require:

$$288 \times 25 \times 352 \times (1 + \tfrac{1}{4} + \tfrac{1}{4}) \times 8 = 30.4 \text{ Mbit/s}$$
and
$$240 \times 30 \times 352 \times (1 + \tfrac{1}{4} + \tfrac{1}{4}) \times 8 = 30.4 \text{ Mbit/s}$$

Although 30.4 Mbit/s is less than 20% of the 162 Mbit/s representing the net capability of studio-quality sources and this reduction has been achieved merely by filtering, this does not mean that 80% of the original quality has been discarded. The resulting 352 picture elements horizontally exceeds the 250 norm of VHS VCRs. With appropriate upsampling for display on conventionally scanned screens, the subjective vertical resolution approaches the actual number of lines, 288 or 240, in the progressively scanned coding format whereas that of the original is only around 360 or 300 because of the Kell factor and interlacing [8]. Although the motion sampling rate has been halved to 25 or 30 Hz, it still exceeds the 24 frames per second of cinematography.

In addition to the above conversion processes the pre-processing stage may include noise reduction by means of a nonlinear temporal filter and the post-processing may include cosmetic filtering which attempts to improve the subjective quality. All pre-processing and post-processing is outside the scope of the standard.

10.5.2 Coding kernel

Ignoring for the moment the reordering buffers the coder is a hybrid arrangement of differential pulse code modulation and DCT. A prediction is subtracted from the input and the resulting prediction error is subjected to a DCT to yield coefficients which are then quantized to a smaller set of possible levels. After variable-length coding to take advantage of the higher probability of occurrence of levels representing prediction errors closer to zero, these codewords are multiplexed with various overhead information and pass through a buffer to the storage device. The instantaneous rate of bit generation depends on the picture complexity and is therefore variable. The buffer performs short-term smoothing to match this to a fixed-rate storage device and can also provide feedback to the quantization process to provide a constant longer-term average rate. The encoder also contains the decoding functions so that it is able to reconstruct the same pictures that the decoder subsequently will and use these to generate the same prediction that the decoder will.

10.5.3 Random access

The importance of random access was reflected in the high weighting factor it received in the ballot process. The requirement is to be able to start, or restart, normal-speed forward playback from any arbitrary point in the encoded material with an acceptably low waiting time. The simplest way to achieve this is to encode each of the pictures, occurring at 25 or 30 times per second, as a self-contained unit without reference to others. This corresponds to the switch in Fig. 10.1 being in position '1'. Such 'intra' mode coding is the basis of the draft JPEG standard [9] but, for equivalent picture quality, typically requires at least three times the number of bits of schemes which exploit the temporal redundancy between the consecutive pictures in motion video. These 'inter' modes of transmitting only picture differences yield substantial bit saving, but cannot support random access because the picture preceding the desired starting one will not have been decoded and will therefore not be available for addition of the difference information.

The solution is to adopt a mix of intra- and inter-coding in which the intra mode is used for one picture every so often and the intervening ones are inter-coded. Random access to any picture can be obtained by starting with the nearest previous intra one and then decoding normally until the desired one is reached. Consequently the maximum waiting time is the time between intra pictures plus the decoding delay through the decoder. A higher proportion of intra pictures gives a shorter maximum wait but at the expense of reducing the picture quality because the average compression ratio is reduced. Accordingly, there is no set figure in the MPEG standard, the choice being open to allow the best trade-off in the particular circumstances since there is no impact on decoders. It is even possible for the ratio of intra to inter pictures to vary within the program. Nevertheless, a common figure is 1 in 12 for the 288-line 25 Hz format, yielding a maximum random access time of somewhat over 0.4 s. The pictures commencing with an intra-coded one and up to the the next intra one are termed a group of pictures.

There is another unfortunate effect of these intra-coded pictures. Because pictures occur at a constant rate but the intra ones generate considerably more bits than the others, the rate of generation is not constant but has peaks. This sets a lower limit on the size of and hence delay through the smoothing buffer. This is at least 0.1 s with the example figures above, and is one reason why the MPEG algorithm in a form with its parameters optimally set for storage applications is not ideally suited to interactive video communications, such as videotelephony, in which the accumulation of excessive end-to-end delay must be combatted.

An alternative to the intra-coding of complete pictures would be to intra-code a small but different part of each succeeding picture such that over the same time interval the entire screen area had been covered. This would eliminate the peaks referred to above and permit smaller buffers. Unfortunately, this approach does not work when motion compensation is used, nor does it offer the possibility to implement fast search modes.

10.5.4 Fast-search modes

In many applications it is convenient to be able to move rapidly forwards or backwards through the program whilst still being able to recognize the content. Such a facility is incorporated in VCRs. The nature of video displays means that it is not possible simply to scale up the speed of operation of all the component parts of the decoder and show all the information in a shorter time. It is also the case that it is difficult or costly to increase the rates of reading the storage medium and of processing the compressed data to reconstruct pictures. Thus an apparent speed-up must be obtained by discarding some of the pictures. The intra pictures can be fully decoded without error even if other data is discarded or not even read from the medium. Thus they are most suitable for implementing fast-search facilities.

It should be noted, however, that the MPEG algorithm by itself does not provide a total solution and consequently does not specify how decoding equipments should provide random access features. For some storage devices, CD-ROM being a prime example, the mechanics of the system do not permit selective reading of part of the medium at high speed. In a CD-ROM drive derived from its audio forefather the disc rotates at constant linear velocity taking between 130 and 300 ms per revolution, depending on radial position. This makes it difficult to arrange the spacing of intra-coded pictures on the spiral to enable rapid jumps between them solely by radial movement of the reading head. Instead, after moving the head, the latency of the disk to rotate until the next intra-coded picture can be read may therefore not be much different from the time which would be taken during normal-speed play. Fast reverse is slightly more demanding.

10.5.5 Motion compensation

Motion compensation is a refinement of picture difference encoding which can yield a further compression gain of about two. Although simple in concept and proposed many years ago, it requires such a massive amount of computation [2] that only recently has it become feasible to implement it in coders operating at real time speed.

Encoding the differences between successive pictures requires no bits to be spent on stationary areas, but moving areas produce differences and

THE MPEG VIDEO CODING ALGORITHM   189

encoding them consumes bits. Objects which move as planar translations can be efficiently encoded as instructions to move them from their positions in the previous picture to new positions before making the difference picture. Only the values of the shifts in the vertical and horizontal directions need be encoded. They are known as motion vectors, and the MPEG algorithm applies them to rectangular picture blocks containing 16 × 16 luminance pels and the spatially corresponding 8 × 8 blocks of each of the colour differences. Whereas in H.261 the vectors have integer-valued components, MPEG extends the resolution to half pel positions, prediction values at these locations being obtained by bilinear interpolation. In many cases the motion is more complex than can be fully described by translation alone and the boundaries of moving objects seldom align with the coding block structure. Nevertheless, motion compensation significantly reduces the magnitude of the picture difference signal and reduces the bit count.

Fig. 10.3   Conventional prediction method.

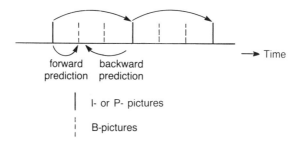

Fig. 10.4   MPEG prediction methods.

The conventional way to incorporate motion compensation is as shown in Fig. 10.3 where the difference picture is formed by subtracting from the picture to be coded the motion-compensated version of the immediately preceding one. The MPEG algorithm is different as illustrated in Fig. 10.4. The first change is that some of the pictures, called P-pictures, are not predicted from the most recent one, but from one further back in time. The second change is that the intervening ones, known as B-pictures, are coded with respect to I- or P-pictures and the motion compensation may be forward from an earlier one or backwards from a subsequent one or a combination of both. In the case where a B-picture is interpolatively coded using earlier and later pictures then two motion vectors per block are required. The possibility to code with respect to two pictures means that MPEG coders and decoders need sufficient memory to hold two complete pictures instead of just one in other algorithms such as H.261. The relationships between the picture types, the coding modes and the positions of the switches in Fig. 10.1 are listed in Table 10.1.

Table 10.1 Possible coding modes for individual blocks in each type or picture

Picture type	Possible coding modes	Switch S1 positions
I	intra	1
P	intra	1
	forward prediction	2
B	intra	1
	forward prediction	2
	backward prediction	3
	forward and backward prediction	4

Because a B-picture cannot in general be encoded until after the next I- or P-picture they are rearranged in order before entering the coding loop. To simplify the decoder the compressed data retains this modified order in which it can be decoded sequentially, the restoration to natural time order taking place just before the post-processing stage. The delay introduced by the reordering depends on the number of contiguous B-pictures. MPEG applications can tolerate the 40 ms per contiguous B-picture. MPEG does not specify the number of contiguous B-pictures as it does not affect the operation of the decoder nor the required capacity of its reordering buffer because only one picture has to be temporarily stored for later insertion at its proper time irrespective of the number of contiguous B-pictures. There is a direct impact on the encoder as the full number of contiguous B-pictures have to be held back. For many image sequences the use of one or two contiguous B-pictures has been found to work well though the standard permits more or none at all.

The delays introduced by the ordering and reordering operations mean that the MPEG algorithm with the use of B-pictures invoked is less optimal than H.261 for two-way audiovisual communication.

B-pictures are never used as the basis for constructing others. It has been found possible to code them with a coarser quantizer than the I- and P- pictures with no visible effect in normal speed play. This can be exploited to gain a further small saving in bits.

The time difference which motion vectors may span in MPEG is larger than in H.261. Also the motion amplitudes in MPEG applications are liable to be much larger than those typically occuring in videotelephony and so the MPEG standard has an extended range of motion vector codewords compared to H.261.

10.5.6 Reverse play

Because the P-pictures employ forward prediction from previous P- and I-pictures, it is not possible to decode MPEG data in reverse to give reverse play. If this feature is required it can be provided by decoding a group of pictures (GOP) in the normal forward manner and buffering them in memory or some other temporary storage device before recalling them in reverse order for display. This procedure would then be repeated for the previous GOP and so on. The number of pictures to be buffered increases with the length of the GOP and they are in their uncompressed form, so short GOPs are preferable. Bearing in mind the random access time limitations of some storage devices, the above reverse-play procedure, with its forward then jump backwards cycles, may not achieve the full speed of normal forward play. Any shortfall will be exacerbated by shorter GOPs where the time to jump backwards represents a larger fraction of the cycle.

10.5.7 Editing of MPEG-encoded video

The MPEG algorithm is primarily intended for delivery of material in its final form. The relatively low bit rate means that the algorithm has to take advantage of the temporal redundancy by using recursive coding. This approach imposes some restrictions on subsequent editing of the video in its coded form. Generally, edits must be of the 'assemble' type, not inserts, and must begin with the intra-coded picture which occurs at the start of a group of pictures. This means that the possible edit points are determined by the original coder and cannot be altered. There is more freedom with the end point of an edit sequence but care must be taken with the buffer fullness so that concatenation with the next edit does not lead to underflow or overflow later.

Some of the restrictions can be avoided by employing intra-coding of every picture and it is conceivable to allocate an equal number of bits to each one so that there is complete freedom to cut and paste at any picture interval. However, these options must be exercised at the encoding stage and will need to be carefully considered against the lower compression factor which can be expected from pure intra-coding.

192 DIGITAL VIDEO CODING

Of course it is possible to decode any MPEG-encoded video back to PCM values, perform any editing (which need not be restricted to cuts but can include cross-fades, manipulation of picture geometry and so on) and encode again, but there will be some quality loss.

10.6. FUTURE MPEG STANDARDS

The above has described the first standard drafted by MPEG. In 1990 MPEG began to move its focus towards the second item of its work plan. The goal of retaining virtually all the subjective quality of CCIR 601 images but compressed to around 10 Mbit/s was found to have widespread support. A key issue is likely to be the level of compatibility or interworking with the first MPEG standard and the means for providing it. Although the activity remains under the ISO with storage in the terms of reference there is considerable interest in, and support for, reaching a solution which will be applicable to digital television broadcasting and telecomunications services. A single standard, or at least a set of very closely related ones, would help prevent the proliferation of incompatible user equipments, and the potentially large manufacturing volumes would bring economies of scale. To that end, liaison arrangements have been established with CCITT and CMTT.

The third item in the MPEG work plan concerns rates up to about 40 Mbit/s and these are likely to prove suitable for high-definition television images (HDTV). MPEG has not yet commenced its studies in this area. It is not certain whether fundamentally different methods will be adopted or if modifications or extensions of the two earlier MPEG standards will be preferable.

10.7. CONCLUSION

The main points of the first MPEG digital video coding algorithm for storage applications have been explained. In reality it is not a rigid set of rules which will produce a single unique compressed file for a given input. The standard defines the syntax and semantics whereby a decoder can interpret the bits it is given to reconstruct the video sequence that the encoder intended. Many parameters in the encoding process are available to optimize the picture quality for the image sequence to be handled or to trade off one feature against another. The standard is not tied to a particular image format nor to a single bit rate or set of bit rates.

The formulation of the MPEG standards has been driven both by technological opportunities and application demands. These standards are eagerly awaited and, by expanding the scope of information technology to include moving images, promise to have a very significant impact on the world.

REFERENCES

1. Sigel E, Schubin M and Merril P F: 'Video Discs: The Technology, the Applications and the Future', Knowledge Industry Publications, Inc. (1980).

2. Whybray M W and Carr M D: 'Practical low bit-rate codecs', Chapter 8 of this book.

3. A C Luther: 'Digital Video in the PC Environment', McGraw-Hill (1988).

4. Hudson G P: 'The international standardization of a still picture compression technique', BT Technol J, $\underline{7}$, No 3 (July 1989).

5. CCITT: 'Video codec for audiovisual services at $p \times 64$ kbit/s', Recommendation H.261, Geneva (1990).

6. ISO/IEC: 'Coding of moving pictures and audio for digital storage media at up to about 1.5 Mbit/s', Committee Draft 11172-1 (1991).

7. CCIR: 'Encoding parameters of digital television for studios', Recommendation 601-2, Düsseldorf (1990).

8. Hsu S C: 'The Kell factor: past and present', SMPTE Journal (February 1986).

9. ISO/IEC: 'Digital compression and coding of continous-tone still images: Part 1: Requirements and guidelines', Committee Draft 10918-1 (1991).

11

MODEL-BASED IMAGE CODING

W J Welsh, S Searby and J B Waite

11.1. INTRODUCTION

When images are coded digitally for storage or transmission, it is desirable to keep the quantity of data generated as low as possible because the cost of storage or transmission is related to this quantity. Compression can be achieved by exploiting the correlation between the pixels of the digitized image. Conventional algorithms used for compression are DPCM, transform coding, quadtree coding and vector quantization [1]. It is possible to achieve some compression without reducing the information content of the image by using variable bit-length codes such as Huffman or arithmetic codes. However, many coding schemes being developed today are intended for a very high compression and this necessitates some degradation of the images. Most of the coding schemes mentioned above can be designed to make these degradations less noticeable to the eye.

Current research effort is being focused on the development of videophone and videoconferencing systems operating over one or two 64 kbit/s channels (speech and video)[2]. The requirement for transmission of video at such a low bit rate stretches current algorithms to their limits and they can exhibit poor performance when there is significant motion in the scene. They may require several frames to be dropped or the spatial resolution to be reduced in order to achieve the compression needed.

The concept of model-based or knowledge-based image coding (MBC, KBC) has arisen because there is a possibility of exploiting much more of the redundancy in a sequence of images than is exploited at present. The term 'model-based' refers to the utilization of models of the objects in the scene. As a simple example, consider a scene containing a cube moving around against a plain background: the information needed to generate a moving sequence at the receiver would be a description of the cube followed by a description of its motion. Of course, the feasibility of extracting redundancy in this way depends on the complexity of the objects and the background. It is fairly straightforward to produce computer-animated sequences of cubes, but for videoconferencing the modelling of people in the scene is required. This chapter will describe work on the modelling of the head and shoulders of a person, for the purpose of transmitting videophone images.

A question which must be considered before proceeding concerns the fidelity of the image displayed. A conventional measure of fidelity is the normalized mean square error (NMSE) between the received image and the original at the transmission end. There is normally a direct relation between this measure and the subjective quality of the image; that is, an image with a high NMSE may appear blurred or contain annoying artefacts. However, it can be argued that if the coded pictures are intended for videoconference or videophony purposes, the important information required is the identity of the subjects, their facial expressions and their approximate bodily movements. As an example, consider a moving sequence in which the subject raises her arm and lowers it again. The requirement for a successful coding scheme would be to display a moving sequence of the woman raising and lowering her arm. If the precise speed, angle and height of the arm displayed were not exactly that which occurred in reality, it might not really matter to the viewer and he would not notice any defect in the image. However, the NMSE between corresponding frames of the original and the displayed sequence might be significant. In a model-based system, the model of the person displayed has to be animated using instructions which have been generated after analysis of the source scene. It is clear that inaccuracies in position may not, in certain circumstances, be crucial. Therefore the NMSE may not be an appropriate measure; the quality of the transmitted image must be judged subjectively.

11.2. APPLICATIONS OF MODEL-BASED IMAGE CODING

Model-based coding can be exploited in two different ways. In the first approach, the images displayed are purely synthetic: a moving sequence could be transmitted by analysing the scene, in order to determine the nature of the objects in it, and then describing the motion of the objects. The information is coded and transmitted to the receiver so that it can be used to set up an internal model of the objects, and their motion coded and transmitted to allow the receiver to produce an animated sequence. Small differences in motion between the reconstructed scene and the original could be allowed if they are below some subjective threshold.

In the second approach, model-based methods are used to form a good prediction for successive frames in a moving sequence [3]. The errors between the prediction and the actual frames can then be coded using conventional image-compression techniques [1].

Predictive techniques are commonly used in image coding systems based on simple methods for compensating the motion. The images are split into a grid of sub-images typically 16×16 pixels in size and each sub-image, or block, in a frame is used to locate the position in the previous frame which is the best match for it. This gives an indication of how each area of the image has moved between frames. A motion vector is obtained for every block and transmitted to the receiver allowing it to set up a prediction frame and this is also set up in the transmitter. It is the difference between this prediction frame and the original frame which is coded and transmitted. The drawback of this sort of scheme is that the motion of the blocks is purely translational and there is likely to be rotation in the image. As the matching is purely local to each block, the 'motion vector' may not actually reflect true motion at all. The outcome of this is that the prediction may be poor in some cases and this can lead to noticeable degradations in the images displayed.

A model-based predictor should give a superior prediction frame allowing the data rate to be reduced without sacrificing picture quality. The trade-off between accurate motion and subjective quality of the images can still be incorporated in this system. In coding the errors, more bits can be allocated to areas of the image which are subjectively important such as the eyes and mouth in a head-and-shoulders videophone image.

The use of model-based prediction holds great promise for the further development and improvement of image coding schemes, and is the subject of current investigation. However, work on the first approach, namely model-based image synthesis, is more advanced and for this reason the chapter will concentrate on this approach.

11.3. REVIEW OF WORK ON MBC

Parke [4] was the first to propose that a parameterized model of a face could be used for a form of videophony. His facial animation is based on a computer model of a head. A common way of representing objects in 3D computer graphics is by a net of interconnected polygons. In this method, the model is stored in the computer as a set of linked lists or linked arrays.

There is a set of arrays giving the *x,y* and *z* co-ordinates of each polygon vertex in object space: $X[V], Y[V], Z[V]$ where V is the vertex address. In addition, a pair of two-dimensional arrays are used: LINV[L][E] gives the addresses of the vertices at the end of a line L, where E is either 0 or 1 depending on which end of the line is being considered: LINL[P][S] gives the line address for each side S of a polygon P. The wire frame is drawn on the screen by iterating through all the values of L in LINV[L][E] giving the vertex addresses which, in turn, yields the co-ordinates of the vertices using the arrays $X[V]$, $Y[V]$ and $Z[V]$. The side of the polygon is projected on to the screen using perspective or orthographic projection. An example is shown in Fig. 11.1.

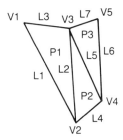

Fig. 11.1 A section of a wire-frame model showing how the vertices, lines and polygons may be numbered. Co-ordinates of vertex 5 are $X[5]$, $Y[5]$, $Z[5]$ indexed by vertex addresses. Addresses of vertices at ends of line 6 are given by LINV[6][0] and LINV[6][1]. Addresses of each side of polygon 3 are given by LINP[3][0], LINP[3][1] and LINP[3][2].

In order to make the models appear more realistic, the polygon net can be shaded. The shading is made to depend on the presence of imaginary light sources in the model world and the assumption of reflectance properties of the model surfaces. If the planar surfaces of the polygons in the wire frame are each given a uniform shading, the faceted appearance becomes apparent. A so-called smooth shading technique such as Gouraud or Phong shading [5] can be used to improve realism. In these techniques the shading in each polygon is effectively interpolated between the shading values at the vertices. This gives the impression of a smoothly curved surface.

In the case of Parke's model, Phong shading was added to improve the appearance. The face can be made to move in a global way by applying a set of rotations and translations to all the vertices as a whole. In addition the vertices of certain polygons can be translated locally in order to change the facial expression. This is in accordance with a system that has been developed for categorizing facial expressions in terms of about 50 independent facial actions (the facial action coding system or FACS) [6]. These actions were translated by Parke into a set of vectors which could be applied to the vertices.

Parke suggested that if the face model could be made to approximate that of a real person then a moving image of that person's face could be generated by determining the motion of the face and using this information to animate the model. By this means, a form of videophony could be established which would require an extremely low amount of information to be transmitted.

Robert Forchheimer and his colleagues at Linkoping University in Sweden have built hardware to generate real-time moving images of a wire-frame face [7]. The facial movements are based on Ekman and Friesen's FACS system. The wire-frame model they use (CANDIDE) has only 100 triangles in order to make real-time moving display possible. A large number of triangles are located in areas of high curvature and in areas which are important for the generation of facial expressions such as the eyes and mouth. A few large triangles are used in flat areas such as the forehead (Fig. 11.2).

Yau of Heriot-Watt University in Edinburgh has worked on a system for modelling a person's head which is based on obtaining a depth map of the head by scanning with collimated laser light [8]. A wire frame is produced by triangulating the depth map using an algorithm which puts small triangles in areas of high curvature and vice versa. A image of the person's face is next overlayed on to the wire frame using a technique called texture mapping [5]. A synthetic view of the person's head can be obtained from any direction by rotating the wire frame and back-projecting the surface detail into an image [9].

11.4. MBC USING FEATURE CODEBOOKS

In this section, a method is described in which a range of facial expressions is generated by a montage process in which sub-images or templates containing facial features are joined together to form a complete face [10]. An algorithm has been developed which automatically builds up codebooks of the templates by training on a moving sequence of images. The synthesis method employs codebooks of eyes and mouths and a single full-face image. With ten eye and ten mouth templates and the single full face, a total of 100 different facial expressions can be produced; the principle is illustrated in Fig. 11.3. Changes are restricted to the eyes and mouth as it is accepted that these are the most important areas concerned with facial expression.

The main problems in implementing this scheme are to locate the eyes and mouth in a face, to derive templates of eye and mouth areas from a head-and-shoulders moving sequence and to select the eye and mouth templates which are the best match for the eye and mouth areas in the sequence. The main objective is to construct a moving sequence in which the movement of the eyes and mouth is convincing approximation to the movement in the original. The way in which these problems have been tackled is described in the following sections.

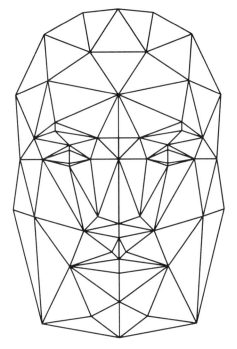

Fig. 11.2 The CANDIDE wire-frame model.

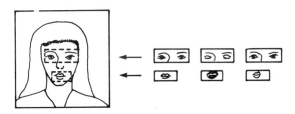

Fig. 11.3 The use of templates of mouths and eyes to create a variety of facial expressions.

11.4.1 Facial image analysis

A facial-recognition algorithm developed by Nagao is employed for the task of locating the eyes and mouth in a face [11]. In this method, a large cross-shaped mask is convolved with the image and the output thresholded to produce a binary image. An original image and processed image are shown in Figs. 11.4 and 11.5; the resolution of these images being 128×128 pixels. The effect of the mask is to low-pass filter the image and apply a Laplacian or second derivative operator. Thresholding the output of this operator is equivalent to isolating negative peaks in the luminance function and produces the subjective effect of a good artist's sketch of a face. This result agrees with the findings of Robinson and Pearson who refer to the line elements in such images as luminance valleys [12]. The pixels which belong to a luminance valley will, in the description of the algorithm, be referred to as valley pixels.

The next stage of the algorithm requires a window to be scanned vertically down the image from top to bottom; the width of the window is the full width of the image and it is 8 lines high. The valley pixels in each column of the window are summed and the values stores in an array. The window is moved down the image by four lines at a time so that the window is overlapped by its previous position. Large values in the array correspond to vertically oriented valleys in the image and as the window is moved down the image two vertical valleys corresponding to the sides of the head or face are located. If two suitable valleys are found, the area between them is searched for symmetrical activity corresponding to the eyes. Locating the eyes enables the axis of the face to be found; a narrow window is then placed centrally on the image extending from the eye line to the bottom of the image. The valley pixels are summed along each row of this window in order to locate the horizontal valleys marking the bottom of the nose, the lips and the chin. The mouth can then be located by a template-matching process. It is possible that a wrong identification of a feature can occur at some stage of the process, and to overcome this problem the original algorithm of Nagao incorporates backtracking so that the process can be restarted from a previous stage and a new candidate selected. Nagao used a large sample of over 700 faces and claims that the inclusion of backtracking improves the performance significantly. However, although he claims a 90% success rate for subjects displayed full-face with no glasses or beard, the performance falls considerably for other cases. The algorithm, in its present form, will only work when the head is vertical. The implementation of the algorithm does not incorporate backtracking and the success rate for location of eyes and mouth is only about 50%. The estimation of the horizontal position of the eyes is good; the algorithm nearly always finds the central axis of the face but the eyebrows are often mistaken for the eyes. Location of the mouth can be confused with the bottom of the nose or the chin.

MBC USING FEATURE CODEBOOKS 201

Fig. 11.4　Original face image.

Fig. 11.5　Laplacian thresholded image.

11.4.2 Feature tracking

The requirement to produce templates of the eye and mouth areas and assemble them into codebooks means that the features have to be tracked very accurately as the head moves. The codebooks must be built up so that the features are always located in the same place in the area covered by the codebook entry. If this were not the case then a noticeable jump might occur in the position of the feature between two frames and a jump in the position of the eyes by only one pixel (in a 128×128 image) is visually disturbing. A similar jump in the position of a mouth is not so disturbing, probably because the mouth is a more flexible object and the exact position is not easy to define.

Initially, Nagao's algorithm was applied to each frame of the sequence in an attempt to locate the features but the positions obtained were too coarse. Therefore, after the features are located in the first frame of the sequence, they are subsequently tracked from frame to frame using a template-matching method applied to a sequence of images processed by the Laplacian operator. This method will be described in terms of the mouth but is equally applicable to the eyes.

The mouth area in the first frame of the sequence is used as a template to find the mouth in the next frame. To find the best position, the template is placed over the image in the same position as it was in the previous frame and a measure of similarity between the template and the current mouth is calculated by summing the number of valley pixels which coincide. The template is next shifted by one pixel to each of the eight surrounding positions and the same measure is calculated. If any of these eight positions returns a higher value than that of the centre position, a search is initiated centred on the higher-scoring position and continues until all eight positions surrounding the centre return a lower score than the centre position which is then determined to be the optimal position for the template.

Of course, the mouth shape may change from frame to frame and this may make the correlation between the template and the current mouth very low and cause tracking to be lost. The solution adopted is to update the template if the score drops below a certain percentage of the previous maximum score; the new template is chosen to be the mouth area in the current frame and a percentage value of 50% is generally used. A question may be raised as to why the template cannot be updated on every frame; in other words why the mouth area in a frame cannot be used as the template for the next frame. Although this was tried, the effect was that the template usually remained fixed in the same position in every frame. This is probably because the head moves so slowly that the effective displacement of the mouth between frames is less than one pixel and if the mouth in the previous frame is used as the template then the highest score obtained will nearly always be at the same position in the next frame. The template, therefore, remains fixed at one position.

11.4.3 Feature codebook generation

For generating the codebooks, the comparisons between mouths are made using the full luminance data of the image. The mouth in the first frame is made the first entry in the codebook and the mouth in the second frame is compared with this first entry by calculating the sum of the absolute differences of the corresponding pixel values. If this value is greater than some arbitrary threshold the mouth shape is considered to be sufficiently different to merit its inclusion into the codebook. This procedure is continued for each frame of the sequence, each time the current mouth being compared with all the mouths in the codebook. The threshold set when creating the codebook determines how many mouths will be put into it; a large threshold will produce few mouths but a sequence coded using the resultant codebook will have rather jerky mouth movements. The technique described is referred to as 'leader clustering' [13].

It is necessary to determine the number of mouths which should be in the codebook. There is evidence that there are only about 13 visually distinct mouth shapes associated with vowel and consonant phonemes [14]. This number must be increased to account for different facial expressions as well as for intermediate positions between the mouth fully open and the mouth closed. The moving sequences used in this work are about five seconds in length (128 frames) and the threshold is set to put about 20 mouths into the codebook. The sequences used to produce the codebooks are then coded using the same codebooks. A synthetic sequence is created by using one full-face image from the original sequence and the best match from the codebook for the mouth in every frame. The best-match mouth is overlaid on the full face in the correct position, the border of the mouth area being blended into the face using a cross-fading technique. The results give an adequate representation of the original mouth movements; however no head movement is introduced at this stage (Figs. 11.6, 11.7).

The most noticeable defect in the coded sequence is the quantization effect due to the small number of mouths in the codebook; one mouth shape tends to be selected for a number of frames and then there is a sudden jump to a different mouth shape which is maintained for a few frames. In order to get over this problem, it is necessary to increase the number of mouths and since the present method requires an exhaustive search of the codebook the time required will increase in proportion to its size.

204 MODEL-BASED IMAGE CODING

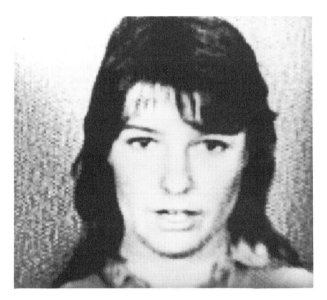

Fig. 11.6 The first of two synthetic images generated by the codebook method.

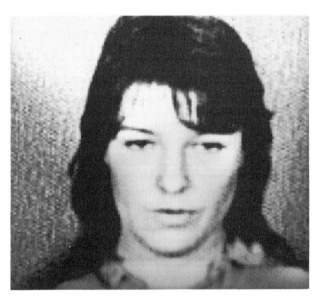

Fig. 11.7 The second of two synthetic images generated by the codebook method.

11.4.4 Data-rate requirements

The codebooks must be derived by analysing a sequence of images of the subject. When communication is established with another user, this data is transmitted down the line to the other user's receiver. If the connection is made over the ordinary analogue phone line, there will be significant delay until all the data is received. For every frame of the moving sequence to be transmitted, the codebook is searched for the best match to the mouth in the current frame and the code for this entry transmitted. The code requires a very small number of bits depending on the number of mouths in the codebook; for a codebook with 64 entries, which is more than is needed for adequate reproduction, six bits are required to code each mouth and a 25 frame/s sequence will only need 150 bits/s to code the mouth movements. The codebook can be doubled for an increase of only one bit per frame but this is at the expense of increased search time and increased time to transmit the codebook data at the start of the transmission.

11.4.5 Disadvantages of the approach

As indicated above there are disadvantages apparent in this method: the features must be tracked very accurately to avoid 'jitter' when the synthetic images are displayed and the codebooks have to be transmitted at a large cost in bits. In the following section, an alternative method for producing facial expressions is described as well as a technique for producing global movements of the head and shoulders.

11.5. 3-D MODELLING OF THE HEAD AND SHOULDERS

A wire-frame model has been developed which can be adapted to the shape of a subject's head and shoulders and a single full-face image of the subject is texture-mapped on to the wire frame. It has been found that the accuracy of the depth values of the wire frame is not critical for producing subjectively acceptable synthetic images as long as the wire frame is not rotated by more than about 30° from the full-face position. The model is an adaptation and extension of the CANDIDE wire frame described in section 11.3. Parts corresponding to the hair, neck and shoulders have been added and the mouth area has been substantiallly refined by the addition of more triangles. A second wire frame is used to model the interior of the mouth (Figs. 11.8, 11.9); a second face image with open mouth is needed to supply image data

206 MODEL-BASED IMAGE CODING

for this extra wire frame and a realistic effect is obtained when the mouth of the main wire frame is opened to reveal the interior of the mouth. Examples are shown in Fig. 11.10.

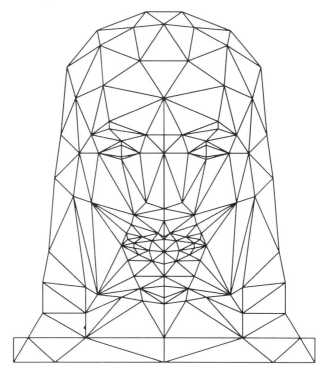

Fig. 11.8 New face model.

Fig. 11.9 Mouth interior model.

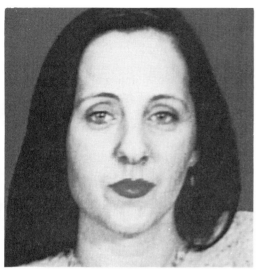
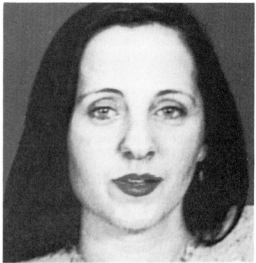
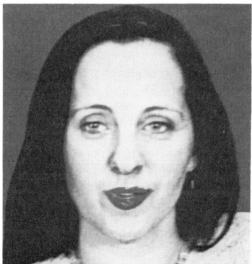

Fig. 11.10   Three examples of a synthetic face image.

11.5.1   Conforming the model to the head and shoulders

In order to conform the model shape to the head and shoulders in the source image, the piecewise linear mapping described by Goshtasby [15] is chosen. Its suitability and method of application are summarised in this section.

The effect of using a piecewise mapping is that the positions of points in the model are only influenced by changes in position of the control points closest to them. Figure 11.11 shows a simple example of the effect of a piecewise mapping when a point is moved.

In Fig. 11.11, the points a, b, c and d are control points and c has been moved away from the others. This has resulted in the features inside the triangle b-c-d being stretched whilst those inside a-b-d are completely unchanged. The positions of non-control points are calculated by interpolating between groups of three control points.

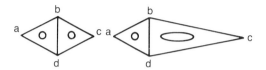

Fig. 11.11 Piecewise linear mapping.

11.5.2 Triangulation of the control points

The criterion for constructing the triangulation of the control points is that any point in a triangle is closer to the three vertices that make the triangle than to vertices of any other triangles. A triangulation of this kind is said to be an optimal triangulation or Delaunay triangulation. There are a number of algorithms describing ways of constructing an optimal triangulation.

The method chosen for use in the current work is due to Sloan [16] and was found to be easily implementable. It is based around the circle criterion which examines pairs of triangles. It works by adding new points, one at a time, into the triangulation and updating it each time to keep it optimal. The first step is to construct a 'supertriangle' which encloses all of the control points. This is the initial Delaunay triangulation. As each new point is added, three new triangles are formed by linking the new point to the vertices of the existing triangle which enclose it. After adding the new point it is necessary to update the triangulation so that it remains a Delaunay triangulation. This is done using Lawson's swapping algorithm [17]. When all the points have been added the original supertriangle is deleted and the desired triangulation has been achieved. Figure 11.12 shows how Lawson's swapping algorithm is used. If P is the newly added point then the circle criterion is applied to the triangle on the edge opposite to P. If P lies inside the circumcircle of that triangle the diagonal V1-V2 is swapped to become P-V3 and the process is repeated for the two triangles which have now become opposite to P. This continues until no more swaps are made.

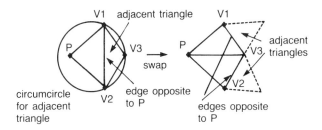

Fig. 11.12 Lawson's swapping algorithm.

11.5.3 Texture mapping and action units

Realism in the synthesized images is achieved by mapping texture from an original picture on to the facets of the model in their new positions. Generating a new image consists of two stages: first the rotation of the wire frame and application of any required facial expressions; second the application of texture to the facets.

Considering first the manipulation of the wire frame: movement of the face involves rotations about the x, y and z axes and changes in facial expressions. Rotations are applied using simple matrix operations. The facial expression changes are based on Ekman and Friesen's system of action units [6]. An action unit consists of a list of vertices of the net together with an associated movement, for each vertex, in x, y and z directions. Actions can be applied to a varying degree by using a multiplying factor in the range 0 to 1. They can also be applied in combination so that, for example, the eyebrows can be raised at the same time as opening the mouth. Once the new co-ordinates of the vertices of the net have been calculated a search is made to produce a list of all the facets which have not been turned over. This can occur when part of the model has rotated out of view and clearly facets which are now viewed from the opposite side should not be shaded. The criterion applied to test the facets is the sign of the vector product of vectors along two sides of the facet. If this sign has been reversed then the facet has flipped over and should not be shaded.

The next step is to sort the remaining facets into order of decreasing depth so that any facets which occlude others will be shaded later. In practice very few facets will cover any others (for small rotations of the head) so this technique is not particularly wasteful in shading areas twice and has the advantage of simplifying processing as it avoids the cost of determining which areas are hidden. The shading process itself takes each facet in turn from the sorted list and maps each pixel in the facet to a corresponding point in the same facet in the original image. To perform this mapping a linear interpolation is used similar to the one used for calculating displacement when conforming the net. The grey-level value for the pixel being drawn is calculated by bilinearly interpolating the values of the four pixels surrounding the point in the original image.

11.6. STATIC IMAGE ANALYSIS — FACE FEATURE LOCATION

11.6.1 Locating the head and shoulders boundary in an image

In order to conform the wire frame to the head and shoulders of a subject, the boundary of the head and shoulders must be located as well as the positions of the internal facial features. An implementation of the 'snake' algorithm of Kass, Witkin and Terzopoulous [18] has been used to find the head boundary [19].

11.6.1.1 The snake algorithm

The snake is an energy-minimizing spline guided by external constraint forces and influenced by image forces that pull it towards features such as lines and edges. In the current work, the image is first convolved using the Laplacian operator of Nagao [9]. The resulting output is modified using the operation

$$f(x) = \frac{x^2}{1 + x^2} \qquad 11.1$$

This sigmoidal function suppresses small levels of activity due to noise as well as very strong edges whilst leaving intermediate values barely changed. By this means, the snake is presented with a smoother energy field which will reduce its tendency to oscillate in position.

The snake is initialized surrounding the image and is then allowed to contract under its own internal elasticity. The Laplacian operator produces energy 'valleys' around the boundary of the head and the snake contracts until it falls within the contour formed by these valleys. If the image forces are great enough the snake is impeded from contracting further and will stabilize. Details of the implementation are given in [19] together with some pictures showing the various stages of contraction of the snake on to a head. In a preliminary experiment using 28 facial images the snake converged successfully in 15 cases. In nine cases, it managed to locate most of the boundary but became trapped at some point by a feature external to the face.

11.6.1.2 Testing the contour shape

Owing to its tendency to get stuck in certain cases, it is essential to determine the shape made by the snake contour after equilibrium is reached. As the snake forms a closed contour, it is possible to decompose the shape function into a set of Fourier components a_n [20].

It has been shown [21] that

$$b_n = \frac{a_{1+n} a_{1-n}}{a_1^2}, \quad n > 1 \qquad 11.2$$

forms a set of coefficients which are invariant with respect to the starting point on the contour as well as to scale and orientation changes. Information about the orientation can be retained by defining a coefficient

$$b_1 = \frac{a_2 |a_1|}{a_1^2} \qquad 11.3$$

The shape descriptors are classified using a multilayer perceptron (MLP), a type of neural network, with eight input nodes, a single hidden layer and one output node [22]. Table 11.1 gives a set of values of the shape descriptors for which the MLP successfully gave the correct classification.

Table 11.1 Fourier descriptors of snake contour

Input data (first eight b coefficients)	Output
0.185 0.459 0.00 698 0.00 189 0.00 168 0.00 615 0.00 136 0.001 140	1
0.660 0.537 0.01 540 0.00 398 0.00 827 0.00 433 0.00 282 0.000 388	0
0.612 0.469 0.00 805 0.00 148 0.00 507 0.00 230 0.00 103 0.000 009	1
0.486 0.323 0.00 170 0.00 061 0.00 062 0.00 123 0.00 023 0.000 148	0
0.449 0.394 0.00 175 0.00 024 0.00 109 0.00 240 0.00 078 0.000 152	1
0.824 0.400 0.00 601 0.00 180 0.01 110 0.00 512 0.00 042 0.000 757	0

An output of 1 indicates head-shaped and 0 indicates non head-shaped.

If the snake's shape is incorrect, it can be reinitialized from a different starting position within the convex hull of the last equilibrium position. Having found the snake to be acceptably head-shaped, the validity of its location can be confirmed by testing for the internal face features such as eyes, nose and mouth. Some methods for finding these features are described in [22]. As the snake is composed of a finite number of nodes, a correspondence can be made between a subset of these nodes and the nodes on the boundary of the wire-frame model in order to conform it automatically.

11.6.2 Internal feature location

11.6.2.1 Introduction

This section discusses work carried out on the location of the internal facial features. This work was carried out on the location of eyes using both template matching [23] and the multilayer perceptron (MLP). An overview of the work is given here and more details can be found in [22]. This apparent success of Baron's approach to face recognition using templates of features encouraged the first experiment. This was followed by a second experiment based on a heuristic technique, the 'symmetrical blob detector'.

11.6.2.2 Symmetrical blob detectors

Symmetry masks

The eye detector is based on Nagao's algorithm which is a model-based approach to face feature location and recognition [11]. This was discussed in Section 11.4.1. Nagao's approach exploits symmetry of the face to aid location of the eyes; the method to be described exploits the symmetry of the eyes themselves. A mask is scanned over the processed image within a predetermined search window (the image having been processed using the Nagao Laplacian operator).

Within the mask, each pixel value is multiplied by a weighting factor which is determined by the pyramidal weighting function shown in Fig. 11.13.

STATIC IMAGE ANALYSIS — FACE FEATURE LOCATION 213

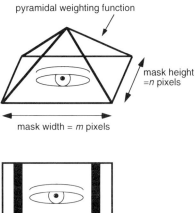

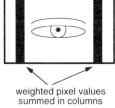

Fig.11.13 Mask for detecting eyes.

This is used to weight the pixels in the centre of the mask area more strongly than those near the periphery so that features which are not part of the eye, but which intrude into the mask area, do not influence the symmetry measure unduly. The weighted values in each column are summed and the total value in each column of the mask which is to the left of the central axis is subtracted from the sum for the equivalent column on the right. The absolute value of these differences are summed to give one of the terms in an overall 'symmetry' measure (clearly, any horizontally symmetrical object will give a zero value for this term although a zero value does not necessarily imply that the object is horizontally symmetrical). A similar term corresponding to vertical symmetry is derived by summing the pixel values along rows. These measures are summarized in the following equations:

$$\text{HSM (horizontal symmetry measure)} = \sum_{i=0}^{m/2} \left| \sum_{j=-n/2}^{n/2} w_{ij} c_{ij} - \sum_{j=-n/2}^{n/2} w_{-ij} c_{-ij} \right| \quad 11.4$$

$$\text{VSM (vertical symmetry measure)} = \sum_{i=0}^{n/2} \left| \sum_{i=-m/2}^{m/2} w_{ij} c_{ij} - \sum_{i=-m/2}^{m/2} w_{i-j} c_{i-j} \right| \quad 11.5$$

where c_{ij} is the value of the pixel in the ith column and jth row and w_{ij} is the weight for that pixel.

In addition, an overall activity measure is obtained by summing the weighted pixel values over the whole mask and a composite measure is determined as follows:

$$\text{SBM (symmetrical blob measure)} = \sum_{i=-m/2}^{m/2} \sum_{j=-n/2}^{n/2} w_{ij} c_{ij} - \alpha \text{HSM} - \beta \text{VSM} \qquad 11.6$$

α and β are constants (currently set to 1) which allow each term to be weighted independently.

Since the feature pixels are added in columns and rows in order to obtain the symmetry measure, it is clear that the detector is also sensitive to skew-symmetric features. In other words, if an eye is centred within the mask but is rotated by an angle of 20° say, the measure should still be the same. This accounts for the rotation invariance mentioned earlier.

The mask is scanned over each eye area and up to ten of the highest local maxima of the SBM function are determined. Local maxima are found by taking the highest value within a small window scanned over the array which contains the values of the SBM function. Although in most cases the overall maximum of the SBM function in each eye area corresponds to the centre of the eye, in some cases it corresponds to the centre of an eyebrow. An example of the peaks found in an eye area is shown in Fig. 11.14; the white dots indicate their position but not the magnitude of the function. In this case, the peak over the eye is over twice the size of any other peak found.

In order to compare the performance of the symmetrical blob detector with that of the MLP and template matching described in [22], an experiment was performed using the same 44 test images as before. Rather than using a fixed search area and fixed mask size, these were dependent on the estimated face size. The estimation of the face size is described in the following section; in this experiment, the two parameters of face size were estimated by manual measurement using a cursor. The two parameters are: the distance between the point midway between the eyes and the chin; and the distance between this point and the side of the face. The results are summarized in Table 11.2.

Table 11.2 Summary of results for symmetrical blob detector used on eyes.

Case	Number of occurrences
highest peak within 2 pels	24
second highest peak within 2 pels	5
highest peak within 6 pels	10
fourth or higher peak within 6 pels	5

STATIC IMAGE ANALYSIS — FACE FEATURE LOCATION   215

Fig. 11.14   Highest peaks in eye area.

Although the results do not appear as good as the best obtained for template matching and the MLP, it is considered that the results would probably be good enough assuming that the technique for confirmation of eye location to be described below were used.

There is more confidence concerning the performance of the symmetrical blob detector on a wider range of images than template matching or the MLP because it is not based on training on a set of images. The images used to train the MLP and template matching in the experiment described in [22] were taken in the same lighting conditions as the images used to test them. Another point is the scale and orientation invariance mentioned earlier.

Setting up search windows for the eyes

Although the work on classifying the snake shape using Fourier descriptors described in section 11.6.2 was showing promising results, there were problems in deciding how to determine a strategy for using the shape information in order to search for internal features. This is important in view

216 MODEL-BASED IMAGE CODING

of the fact that the snake might assume a perfectly valid head shape whilst trapped in the wrong position such as on the subject's collar. Only a directed search for the eyes and mouth seemed a possible way of determining the status of the snake. Fitting an eclipse to the snake contour provides a straightforward way of setting up search windows for the eyes and mouth.

The original method of testing for snake equilibrium measures the overall distance moved by each node of the snake over five iterations. If this is less than an arbitrary threshold, the snake is judged to have stabilized; after this, the parameters of the best-fit ellipse to the snake contour are found by least squares. This method works by minimizing the sum of the squared distances between the snake nodes and the ellipse. The parameter values determined are: the x and y co-ordinates of the centre; the lengths of major and minor axes; and the orientation. The major and minor axes are displayed on the screen. Results obtained with this method are good (see Fig. 11.15 which shows the major and minor axes of the ellipse superimposed on the snake in equilibrium).

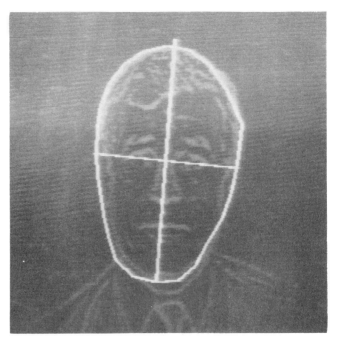

Fig. 11.15 Fitting an ellipse to the snake contour.

It is found that, for the 15 cases in which the snake successfully locates the head boundary, the minor axis of the fitted ellipse is close to the line joining the eyes of the subject. Consequently, a search window is set up with its height and width proportional to the minor axis length. The proportions of the window are shown in Fig. 11.16.

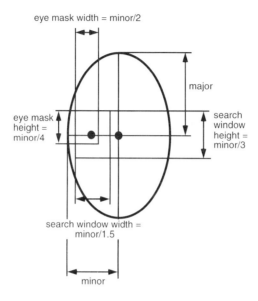

Fig. 11.16 Proportions of search window and mask for an eye detector.

Nose and mouth location: confirmation of eye location

It was stated above that the eye locator sometimes finds a maximum on an eyebrow rather than on an eye. In order to distinguish between the two cases a further stage is used to provide more evidence for one case or the other. Successive pairs of peaks are taken, one from each eye area, the highest values being considered first. The distance between the peaks is measured and compared against a high and low threshold derived from the length of the minor axis of the fitted ellipse. If this distance is within an acceptable range the next stage is carried out; otherwise, another pair of peaks is considered. In the following stage, another search window is set up as shown in Fig. 11.17. Within this window, a mask similar to that described in the section on symmetry masks is used to find a set of peaks. As shown in Fig. 11.18 the peaks tend to line up along the axis of symmetry through the nose and mouth; the angles between the lines joining these peaks to the midpoint of the line

218 MODEL-BASED IMAGE CODING

joining the eyes and the line joining the eyes are used to determine the likelihood of a correct configuration. If the mean value of these angles is close to 90° and the variance of the angles is small then a good indication is given that the pair of peaks which are candidates for giving the location of the eyes do indeed correspond to the eyes. Figure 11.19 shows a case where an invalid configuration of eye candidate points and nose-mouth axis is rejected. If the highest peaks were detected on each of the eyebrows, this would probably be accepted as a valid configuration. Such an error could be avoided by continuing to search for a pair of peaks of smaller value that satisfy the above geometrical constraints but which are closest to the bottom edge of the image.

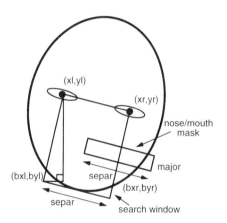

Fig. 11.17 Search window for nose and mouth.

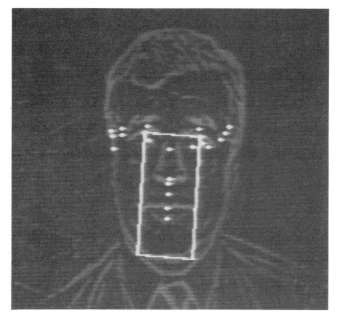

Fig. 11.18 Peaks found in nose-mouth area.

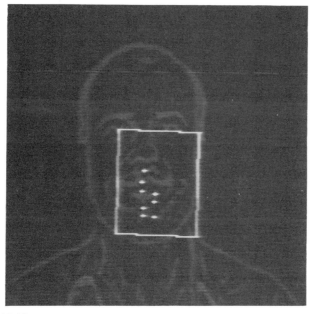

Fig. 11.19 Invalid configuration of eye candidates and nose-mouth axis.

11.6.3 The 'snake-segments' method

The snake is capable of locating head boundaries in most of the cases tested so far but there are cases where information may be required which cannot be found using the snake in the straightforward ways discussed in the previous sections. For example, in some cases the conformation of the wire frame requires the location of the hair-face boundary. It is not possible to locate this boundary by contracting a closed-loop snake on to it as before since it would be trapped by the hair-background boundary. In addition, it is required to locate features such as the boundary between the shoulders and the background. A more general type of method was developed for finding these features and this is described with reference to Fig. 11.20.

A large set of fixed-end snakes are fitted to boundary segments in an image. The fixed ends are located by dividing the image into a number of horizontal bands and searching for peaks in the Nagao processed image along the lines separating the bands. In each of the bands, a number of snakes are successively initialized between the 12 highest peaks on the line at the top end of the band and each of the 12 highest peaks lying on the line at the bottom of the band which lie within some horizontal distance threshold. Each snake is allowed to iterate a few times after which, if the two end points lie on the same piece of boundary, the snake will have settled along the length of the boundary. Otherwise it will assume some arbitrary path between the two points. It is possible to determine which of the two cases applies by summing all the pixel values of the Nagao processed image lying beneath the snake; if it is on a boundary this sum will be higher than if not.

After repeating the process for each snake in the band, the positions of the eight snakes which have the highest level of featural activity are recorded and then the next band is similarly processed. Eight snakes are chosen in each band as this gives a reasonable chance of building concatenations corresponding to the sides of the head and (or) the sides of the face. After repeating this for each of the six bands, a total of 48 snake segments will have been stored. The next stage is to link up the snakes into concatenations of snake segments. To achieve this, a list of concatenations is generated which is updated as each band is processed moving from top to bottom of the image. First, the eight snakes in the topmost band of the image (band 1) are put into the list as concatenations of length 1. Following this the snakes in the next band down (band 2) are tested to determine if they connect to the bottom end of any band 1 snakes. Depending on the results, some of the concatenations previously recorded may be modified to become length 2; new length 2 concatenations may be added or, if the band 2 snakes do not connect to the band 1 snakes, some new length 1 concatenations are added. After finding all the possible concatenations, their featural activity is measured and stored.

STATIC IMAGE ANALYSIS — FACE FEATURE LOCATION 221

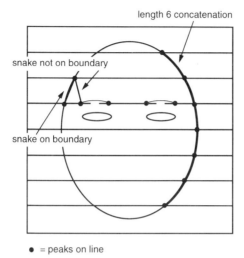

Fig. 11.20 The snake-segments technique.

Fig. 11.21 Hair-face boundaries located using snake segments.

Following this, a process of elimination is carried out which deletes all the concatenations of a given length which lie in roughly the same position except for the highest activity one. A notional position is assigned to each concatenation which is the centroid of the end points of the composite snake segments. The deletion process reduces the number of concatenations drastically. Generally, it is found that the longest and highest activity concatenations correspond to important features such as the hair-face boundaries or the hair-background boundary. Figure 11.21 shows the two longest and highest activity concatenations found.

11.7. GLOBAL MOTION ESTIMATION

In the MBC technique, the head-and-shoulders model and facial expression changes are transmitted to the receiver but global motion is also required. The model has to be animated using instructions which have been generated after analysis of the motion of the person in the source image. A method has been developed using local motion estimation to analyse translations and rotations of a subject's head in order to synthesize a realistic moving head matching the original motion. The method used is discussed in this section.

11.7.1 Local motion estimation and geometrical model

Consider a moving image sequence; a set of points is used to estimate the motion from one image to the next. A set of local motion estimates is obtained using best match block search. Consider a block $b1*b1$ around a point $p1$ in the first image and another block $b2*b2$ ($b2 > b1$) around a point with the same co-ordinates in the next image. The best match block search consists in testing all the blocks $b1*b1$ in the block $b2*b2$ and adding the absolute value of the differences between the pixels of the $b1*b1$ block of the first image and the $b1*b1$ block of the second image. The minimum corresponds to the best fit of the block in the new image.

Figure 11.2.2 shows the best match block search. If dx and dy are the displacements of the point $p1$ between image 1 and image 2 then

$$dx = x'-x \qquad\qquad 11.7$$

$$dy = y'-y \qquad\qquad 11.8$$

In general the motion can be due to a combination of rotation and translation. Considering a moving point (x,y,z); X and Y displacements are decomposed into X,Y translations and X,Y,Z rotations.

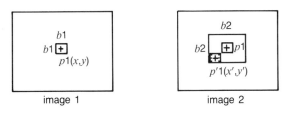

Fig. 11.22 The best-match block search.

It is assumed that the camera is far enough away that perspective effects should not make any great contribution and, therefore, Z displacements are not considered. Also, the Y axes in the geometric and screen frames of reference are opposite in sense as shown in Fig. 11.23.

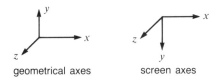

Fig. 11.23 Relation between geometric and screen frames of reference.

The rotation procedures are implemented according to the geometrical axes and angles which are positive for clockwise rotation. Therefore to obtain consistency, X rotations must be positive when anticlockwise and Y and Z rotations positive when clockwise.

X displacements are a consequence of X translation, Y and Z rotations as shown in Fig. 11.24.

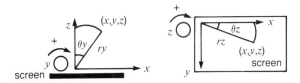

Fig. 11.24 Effects contributing to X displacements.

Y rotation: Z rotation:

$\tan \theta_y = x/z$ 11.9 $\tan \theta_z = y/x$ 11.10

$ry = (x^2 + z^2)^{1/2}$ 11.11 $rz = (x^2 + y^2)^{1/2}$ 11.12

$dx = ry.\cos\theta_y.d\theta_z + rz.\cos\theta_z.d\theta_z + dT_x$ 11.13

224 MODEL-BASED IMAGE CODING

Y displacements are a consequence of Y translation, X and Z rotations, as shown in Fig. 11.25.

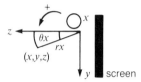

Fig. 11.25 Y displacement due to rotation around X axis.

$$\tan \theta_x = y/z \qquad 11.14$$

$$rx = (y^2 + z^2)^{1/2} \qquad 11.15$$

$$dy = rx.\cos\theta_x.d\theta_x + rz.\cos\theta_z.d\theta_z + dT_y \qquad 11.16$$

In the above cases, the centre of rotation is considered to be the origin (0,0,0).

11.7.2 Least squares estimation

In order to obtain accurate estimates, a large set of moving points are used to obtain local motion vectors and the 'pseudo-inverse method of least squares estimation' [23] is used to discover the best-fitting values of the global motion parameters. In fitting an $n*1$ observations matrix Y by some linear model of p parameters, the prediction is that the linear model will approximate the actual data. Then

$$Y = XB + E \qquad 11.17$$

where X is an $n*p$ formal independent variable matrix, B is a $p*1$ parameter matrix whose values are to be determined, and E represents the difference between the prediction and the actuality: it is an $n*1$ error matrix. For this case, a null error matrix is assumed. Using n points to estimate the motion, $Y_h = X_h.B_h$ is calculated for the horizontal displacements where:

$$Y_h = \begin{Bmatrix} dx_1 \\ . \\ . \\ . \\ dx_n \end{Bmatrix} \qquad 11.18$$

$$X_h = \begin{Bmatrix} 1 & ry_1.\cos\theta_{y1} & rz_1.\cos\theta_{z1} \\ & \cdot & \\ & \cdot & \\ & \cdot & \\ & ry_n.\cos\theta_{yn} & rz_n.\cos\theta_{zn} \end{Bmatrix} \qquad 11.19$$

$$B_h = \begin{Bmatrix} dT_x \\ d\theta_y \\ d\theta_z \end{Bmatrix} \qquad 11.20$$

and $Y_v = X_v.B_v$ for the vertical displacements where:

$$Y_v = \begin{Bmatrix} dy_1 \\ \cdot \\ \cdot \\ \cdot \\ dy_n \end{Bmatrix} \qquad 11.21$$

$$X_v = \begin{Bmatrix} 1 & rx_1.\cos\theta_{x1} & rz_1.\cos\theta_{z1} \\ & \cdot & \\ & \cdot & \\ & \cdot & \\ 1 & rx_n.\cos\theta_{xn} & rz_n.\cos\theta_{zn} \end{Bmatrix} \qquad 11.22$$

$$B_v = \begin{Bmatrix} dT_y \\ d\theta_x \\ d\theta_z \end{Bmatrix} \qquad 11.23$$

The task is to find B_h and B_v with

$$B = (X^T.X^{-1}).X^T.Y \qquad 11.24$$

So the results of the motion estimation are B_v and B_h components: dT_x, dT_y, $d\theta_x$, $d\theta_y$, $d\theta_z$ (which is the average between the two $d\theta_z$ calculated according to the horizontal and vertical displacements).

11.7.3 Tests with a synthetic sequence

For the method to be accurate it is necessary to have precise x,y,z co-ordinates for the points at which the local motion estimates are performed. The global motion estimation was initially tested using a synthetic head sequence as the source as this guarantees correct values for the coordinates. The algorithm was tested as follows.

- A synthetic moving head sequence was created.
- A search was made for a set of corresponding points between the first image and the following one.
- Estimates were made of the global motion parameters.
- The estimated translations and rotations were applied to the wire-frame model with the head and shoulders of the first image texture-mapped on to it in order to produce a prediction of the following image.
- The next search was carried out between the newly synthesized image and the actual image in the following frame.

The algorithm was tested using different basic motions and then with combinations of these. The results for the case of rotation around the x axis are shown in Table 11.3. This shows some spurious y translation as might be expected but as this is at most only about 0.25 of a pixel it can safely be ignored.

Table 11.3 Results of X axis rotation test

Actual motion:

dTx (pixels)	dTy (pixels)	dOx (radian)	dOy (radian)	dOz (radian)
0	0	2×10^{-2}	0	0
0	0	-2×10^{-2}	0	0
0	0	-1×10^{-2}	0	0
0	0	-4×10^{-2}	0	0
0	0	-1.2×10^{-2}	0	0
0	0	-2×10^{-3}	0	0

Estimated motion:

dTx (pixels)	dTy (pixels)	dOx (radian)	dOy (radian)	dOz (radian)
-3×10^{-8}	-2.05×10^{-1}	2.46×10^{-2}	0.85×10^{-9}	3.88×10^{-9}
-9.55×10^{-7}	2.37×10^{-1}	-1.58×10^{-2}	2.08×10^{-8}	6.67×10^{-9}
3.49×10^{-8}	9.44×10^{-2}	-1.47×10^{-2}	4.99×10^{-9}	0.59×10^{-9}
4.70×10^{-8}	1.56×10^{-1}	-3.10×10^{-2}	-1.38×10^{-9}	-5.62×10^{-9}
-8.72×10^{-7}	4.16×10^{-2}	-1.66×10^{-2}	1.72×10^{-8}	6.61×10^{-9}
3.53×10^{-8}	-4.67×10^{-7}	9.90×10^{-2}	-0.93×10^{-9}	-5.17×10^{-9}

11.7.4 Tests with an original sequence

Testing the method with an original sequence produces some problems not encountered with a synthetic sequence:

- the depth values of the synthesized head are not consistent with the depth values of the actual head;

- the origin (0,0,0) is the centre of rotation of the synthetic moving head but not of the actual moving head.

When the net is conformed to the front-facing head-and-shoulders image, the X and Y values of vertices are changed so that they match the X and Y values of corresponding points but the Z values are not changed. A conformation of the net to the profile is necessary in order to adjust the Z values of points to be close to the actual values. The centre of rotation was manually estimated to be on the mid-axis of the face behind the mouth on the front view of the face and just below the ear on the side view. In order to reduce spurious rotations, estimates of rotation which were less than 0.01 radian were ignored.

The method was tested on a sequence of a real subject producing a range of different basic and combined motions in the same way as with the synthetic sequences. When the centre of rotation is well chosen, the results are quite good. The global motion is reproduced in the resulting synthetic moving head but some spurious translations and rotations are sometimes added. Examples are shown in Figs. 11.26 – 11.28.

228 MODEL-BASED IMAGE CODING

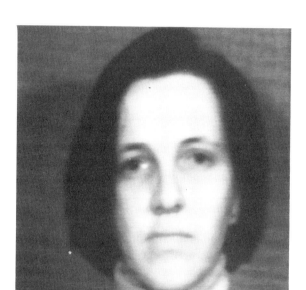

Fig. 11.26   Original head position.

Fig. 11.27   Head position after a few frames.

AUTOMATIC CONFORMATION OF THE WIRE-FRAME MODEL

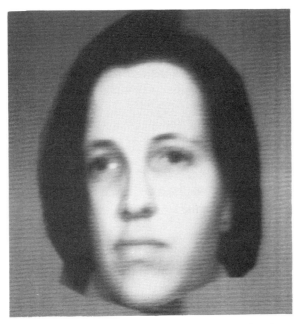

Fig. 11.28   Estimated head position according to the motion estimation technique used.

11.8. AUTOMATIC CONFORMATION OF THE WIRE-FRAME MODEL

In order to conform the wire-frame model, the positions of a set of control points must be automatically determined. The control points can be divided into a number of sets depending on which features of the subject they correspond to. Considering Fig. 11.8, it is apparent that some points correspond to the boundaries between the head and the background, the face outline, the shoulders and the internal facial features. In the work presented so far the automatic analysis can only locate a subset of these points; the shoulders are not located and the internal face features are only coarsely located.

The methods discussed in sections 11.6.1, 11.6.2 and 11.6.3 are used to locate these points; a command program is used to call each of the respective programs in turn and the coordinate values of the points are stored in data files as the analysis progresses. Finally, a program is called which used the co-ordinates in these files to conform the wire frame and display it overlaid on the image of the subject.

230 MODEL-BASED IMAGE CODING

After the head-and-shoulders image is read into the frame store it is processed by the Nagao operator as described in section 11.6.1. Then a fixed-end snake is collapsed on to the head boundary, the final coordinates of the snake being stored in a file. The snake segments technique discussed in section 11.6.3 is then used. First of all the program reads in the file stored after using the fixed-end snake, this being used to estimate the position of the vertical axis of the face and to locate the positions of the left and right halves of the fixed-end snake. After the program runs, the two longest concatenations of snake segments are chosen which are:

- sufficiently distant from the vertical axis of the face and on opposite sides to it;

- sufficiently distant from the left and right halves of the fixed-end snake and fall within it.

This is illustrated in Fig. 11.21.

If suitable concatenations are found, the co-ordinates of the end points of their segments are stored in a file together with the lengths of the concatenations.

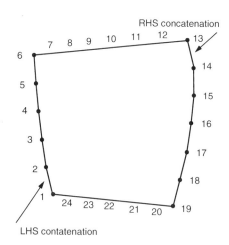

Fig. 11.29 Initialization of closed-loop snake using results of snake segments program.

In the next stage, a closed-loop snake is used to find the face outline. This snake can be initialized in two different ways depending on the results of the snake-segments program. If two suitable concatenations are found, the snake is initialized as shown in Fig. 11.29.

As shown in Fig. 11.29 the concatenation on the left has five segments and so it is used to initialize the positions of the first six nodes of the snake. The concatenation on the right has six segments and so initializes the positions of a further seven of the snake nodes. The remaining 11 snake nodes are positioned by joining the top and bottom ends of the concatenations with straight lines, as shown, and distributing the remaining nodes evenly along these lines. The snake is then allowed to iterate with a fixed number of iterations (100 is sufficient to allow stabilization) using image forces which act so as to attract the snake into the valleys formed around the face outline. The initial and final positions for the snake in one particular case are shown in Figs. 11.30 and 11.31.

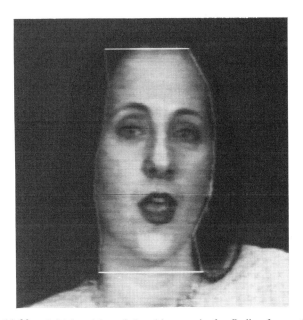

Fig. 11.30   Initial position of closed-loop snake for finding face outlines.

232   MODEL-BASED IMAGE CODING

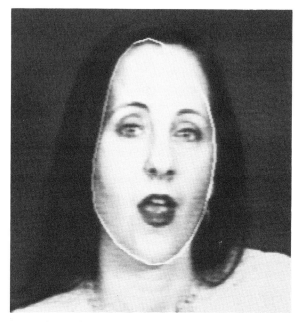

Fig. 11.31   Final position of closed-loop snake.

In the case where two suitable concatenations are not found, the snake is initialized as shown in Fig. 11.32 using the positions of the nodes of the fixed-end snake previously fitted to the hair-background boundary.

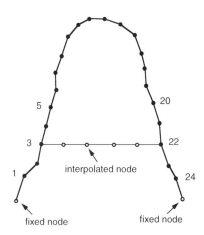

Fig. 11.32   Initialization of looped snake where two distinct concatenations are not found.

AUTOMATIC CONFORMATION OF THE WIRE-FRAME MODEL   233

The mapping between looped-snake nodes and fixed-end snake nodes is given in Table 11.4.

Table 11.4   Mapping between looped-snake nodes and fixed-end snake nodes

Looped nodes	Fixed-end nodes
1	3
2	4
3	5
4	6
5	7
6	8
7	9
8	10
9	11
10	12
11	13
12	14
13	15
14	16
15	17
16	18
17	19
18	20
19	21
20	22

The positions of the looped-snake nodes 21 to 24 are given by linearly interpolating between the positions of fixed-end snake nodes 3 and 22. In both cases, the final positions of the closed-loop snake are written into a file. The search for eyes, nose and mouth is performed as described in section 11.6.2.2 after first fitting an ellipse to the looped-snake contour in order to scale and position the search windows. As the nose-mouth search produces a single set of peaks, it is necessary to try to interpret the positions of these peaks in order to locate the nose and mouth independently. As the peaks tend to cluster into two sets in each of the nose and mouth areas, the positions of the clusters are located according to the method shown in Fig. 11.33.

The position of the peak which is vertically the highest up in the image is determined as well as the vertically lowest peak; then the midpoint of the two points is found. The assumption is made that all the peaks lying above the midpoint belong to the cluster corresponding to the nose, while all those lying below the midpoint belong to the cluster corresponding to the mouth. The centroids of the two clusters are then determined in order to give single positions for these features. Finally, the coordinates of the points corresponding to the eyes, nose and mouth are written into another file. It is appreciated that this method could be easily refined or improved upon using techniques related to those discussed in section 11.6 for eye location;

however, this simple approach has been adopted for the purpose of having a proof-of-concept demonstrator.

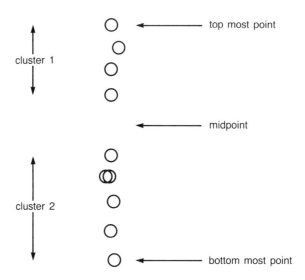

Fig. 11.33   Clustering of peaks in the nose and mouth areas.

The program to conform the wire-frame model is then invoked. First, the wire frame is vertically scaled according to the dimensions of the stablished fixed-end snake. The scaling factor is given by the following formula

$$\text{scaling factor} = \frac{y_s - y_b}{y_{hv} - y_{lv}}$$

where

y_s = y co-ordinate of snake node nearest the top of the image

y_b = y co-ordinate of snake node nearest the bottom of the image

y_{hv} = y co-ordinate of highest vertex of wire-frame

y_{lv} = y co-ordinate of lowest vertex of wire-frame

AUTOMATIC CONFORMATION OF THE WIRE-FRAME MODEL   235

A mapping is set up between the vertices of the wire-frame and the co-ordinates of the points stored in the file. This mapping is given in Table 11.5, the vertex numbers being illustrated by the simplified representation of the wire-frame shown in Fig. 11.34.

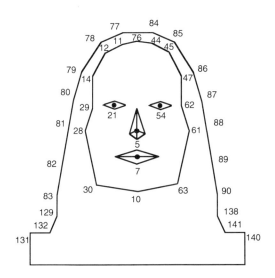

Fig. 11.34   Simplified representation of wire frame showing vertex numbers.

In Table 11.5, os *n* and is *n* refer to outer snake node *n* and inner snake node *n*, respectively. The vertices corresponding to the shoulders are fixed in the image.

Table 11.5 Mapping between wire-frame vertices and located points

Vertex number	Located point
129	os 1
83	os 2
82	os 3
81	os 5
80	os 6
79	os 8
78	os 10
77	os 12
84	os 13
85	os 15
86	os 17
87	os 19
88	os 20
89	os 22
90	os 23
138	os 24
30	is 1
28	is 4
29	is 5
14	is 6
12	is 7
11	is 8
76	is 10
44	is 12
45	is 13
47	is 14
62	is 15
61	is 16
63	is 19
94	is 20
93	is 21
10	is 22
92	is 23
91	is 24
21	left side eye
54	right side eye
5	nose
7	mouth

The initial and final positions of the generic wire-frame model are shown for one case in Figs. 11.35 and 11.36.

AUTOMATIC CONFORMATION OF THE WIRE-FRAME MODEL

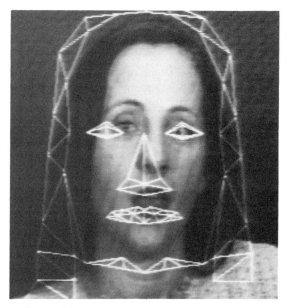

Fig. 11.35 Initial position of generic wire frame.

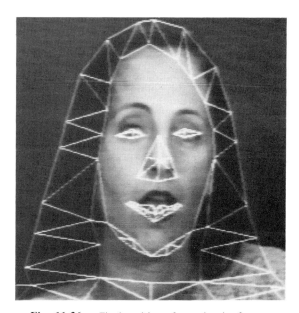

Fig. 11.36 Final position of generic wire frame.

238 MODEL-BASED IMAGE CODING

Another case is shown in Fig. 11.37; in this case the subject has short hair and the true head boundary is given by the equilibrium position of the closed-loop snake. A similar result would be obtained if a closed-loop snake had been used from the outset as described in section 11.6.1. However, note that the closed-loop snake used in the manner described in the current section is initialized closer to the head boundary than in the original experiment; this means that there is a reduced tendency for the snake to become trapped on features such as the subject's collar. This obviates, to some extent, the necessity to resort to the procedures described in section 11.6.2.

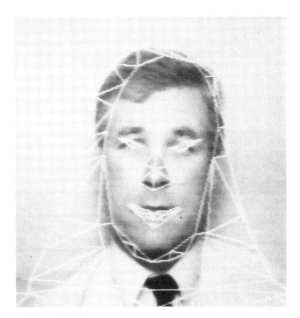

Fig. 11.37 Final position of generic wire frame.

The cases shown in Figs. 11.36 and 11.37 have been treated as successes since all the important features have been located. For comparison, a case where some but not all features have been located is shown in Fig. 11.38. Figure 11.39 shows a case where the hair-background boundary location has failed resulting in failure of the rest of the location.

AUTOMATIC CONFORMATION OF THE WIRE-FRAME MODEL

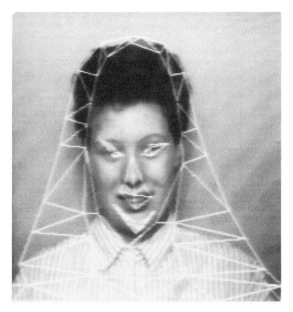

Fig. 11.38 Partial success of automatic conformation.

Fig. 11.39 Failure of automatic conformation.

The results are summarized for a sample of 16 randomly selected images. In Table 11.6, C indicates a complete success, P a partial success and F a failure according to the criteria given above.

Table 11.6 Results of automatic wire frame conformation

Image	Result
DOMIN (Fig. 31)	C
a03	F
a05 (Fig 32)	C
a06	P
a07	P
a11	C
a12 (Fig. 33)	P
a14 (Fig. 34)	F
a20	P
a28	F
a32	P
a38	P
a41	C
a46	P
a58	P
a60	C

Figures 11.40 shows two frames of a synthetic moving sequence generated after the automatic conformation of the wire frame. In this sequence, arbitrary rotations were applied to the model.

AUTOMATIC CONFORMATION OF THE WIRE-FRAME MODEL

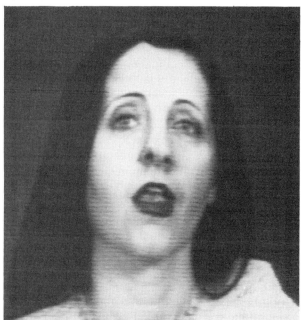

Fig. 11.40 Two frames of a synthesized moving sequence.

11.9. CONCLUSIONS

This chapter has discussed the concept of model-based coding in which a synthetic image of a person's head and shoulders is used to approximate the actual image at the transmission end. Two alternative methods for generating synthetic face images were introduced: the codebook method and the action units method. The codebook method has some disadvantages.

- The codebooks have to be formed off-line and transmitted to the receiver at the beginning of the transmission. This is expensive in terms of the number of bits required and there would be a significant delay before moving pictures could be displayed. The codebooks have to be created by training on a suitably large number of images of the subject.

- Unless the face features are tracked very accurately during formation of the codebooks, the face features will not always be at the same relative position in the codebook window. This leads to 'jitter' when codebooks entries are selected and displayed.

An implementation of an adaptive contour model or 'snake' was found to be a fairly robust method for locating the boundary of a head in an image. The location of the head boundary was used to guide the location of the eyes and mouth; a kind of symmetry detector was developed for this purpose. The snake was put to further use in the 'snake segments' method in which short fixed-end snakes are distributed in the image in order to locate other important boundaries such as the hair-face boundary. Global motion estimation has been implemented and reasonable results obtained using block motion estimation and the least-squares method. Further work is required to determine the motion of the internal face features.

Finally, it was shown how the automatic location of the head boundary and the internal face features can be used to conform the wire-frame model.

Further work is required to develop means to determine the fine structure of the internal face features in order to improve the conformance of the wire-frame model. It is conceivable that, if these improvements are implemented, the analysis and synthesis components could be integrated into a practical system operating in the region of 1 kbit/s.

ACKNOWLEDGEMENTS

The authors wish to acknowledge the contributions to the work described in this chapter of Julie Culver (BTL), Phil Picton (Open University) and Elisabeth Brigant (INSA, Rennes, France).

REFERENCES

1. Pratt W K: 'Digital image processing', John Wiley & Sons (1978).

2. Whybray M W and Carr M D: 'Practical low bit-rate codecs', Chapter 8 of this book.

3. Musmann H G, Hotter M, Ostermann J: 'Object-oriented analysis-synthesis coding of moving images', Image Communication 1, 117—138 (October 1989).

4. Parke F I: 'Parameterised models for facial animation', IEEE Computer Graphics and Applications, 12 (November 1982).

5. Newman W M, Sproull R F: 'Principles of interactive computer graphics', McGraw-Hill (1981).

6. Ekman P, Friesen W V: 'Manual for the facial action coding system', Consulting Psychologists Press, Palo Alto, CA (1977).

7. Rydfalk M: 'Candide, a parameterised face', Research Report of the Dept of Electrical Engineering, Linkoping University, Sweden.

8. Duffy N D, Yau J F S: 'Facial image reconstruction and manipulation from measurements obtained using a structured lighting technique', Pattern Recognition Letters, 7, pp 239—243 (April 1988).

9. Yau J F S, Duffy N D: 'A texture mapping approach to 3-D facial image synthesis', 6th Annual EUROGRAPHICS (UK) Conference, Sussex (6—8 April 1988).

10. Welsh W J, Fenn B A, Challener P: 'Image Encoding and Synthesis', US Patent No. 4841575 (20 June 1989).

11. Nagao M: 'Picture recognition and data structure', from Graphic Languages, ed Nake, Rosenfeld — North Holland (1982).

12. Pearson D E, Robinson J A: 'Visual communication at very low data rates', Proc IEEE, 73, pp 795—812 (April 1985).

13. Hartigan J A: 'Clustering algorithms', John Wiley & Sons (1975).

14. Berger K W: 'Speechreading: principles and methods', Baltimore, National Education Press, pp 73—107 (1972).

15. Goshtasby A: 'Piecewise linear mapping functions for image registration', Pattern Recognition, 19, No 6, pp 459—473, (1978).

16. Sloan S W: 'A fast algorithm for constructing Delaunay triangulations in the plane', Adv Eng Software, 9, No 1, pp 34—35 (1987).

17. Lawson C L: 'Software for C interpolation', Mathematical Software III, ed J Rice, Academic Press, New York, pp 161—194 (1977).

18. Kass M, Witkin A, Terzopoulos D: 'Snakes: active contour models', Int J Comp Vision, pp 321—331 (1988).

19. Waite J B and Welsh W J: 'Head boundary location using snakes', Chapter 12 of this book.

20. Zahn C T, Roskie R Z: 'Fourier descriptors for plane closed curves', IEEE Trans Comp C—21, pp 269—281 (1972).

21. Grandlund G H: 'Fourier preprocessing for hand print character recognition', IEEE Trans Comp, C—21, pp 195—201 (1972).

22. Nightingale C, Myers D H, Vincent J M and Hutchinson R A: 'Artificial neural nets for image processing in visual telecomms: a review and case study', Chapter 14 of this book.

12

HEAD BOUNDARY LOCATION USING SNAKES

J B Waite and W J Welsh

12.1. EDGE DETECTION

A classical problem in image processing is the detection, location and description of the edges or boundaries of objects in an image. Classical edge-detection algorithms provide information on edges in the form of an 'edge image' [1]. In general they work by setting the grey level of each pixel in the 'edge image' to a value that is dependent on the magnitude of the gradient of that the edge detector does not explicitly possess. However, when an edge-detected image is viewed the edges are clearly visible. This is because our visual system is able to 'post-process' the image and provide us with the continuity information that is inherent in edges. Note that this information is present in the edge image even though the detector does not extract it.

For conformation of the wire-frame model in model-based coding [2] it is necessary to find the boundary of the head. Traditional edge detectors in general will not form an edge that is completely closed (i.e. forms a loop around the head) but will instead create a number of edge segments that taken together outline the boundary of the head. A method is required that processes the edge segments and generates a smooth extension that describes the head boundary. Snakes appear to show promise in this direction. The remainder of the chapter describes the theory of snakes and shows in detail how they may be implemented. Section 12.8 shows some results of applying snakes to the problem of finding head boundaries.

12.2. SNAKES

Introduced by Kass *et al.* [3], snakes are a method of attempting to provide some of the post-processing that this visual system performs. A snake has built into it various properties that are associated with both edges and the human visual system (e.g. continuity, smoothness and to some extent the capability to fill in sections of an edge that have been occluded).

A snake is a continuous curve (possibly closed) that attempts to dynamically[1] position itself from a given starting position in such a way that it 'clings' to edges in the image. The form of snake that will be considered here consists of curves that are piecewise polynomial. That is, the curve is in general constructed from N segments $\{x_i(s), y_i(s)\}, i = 1, \ldots, N$, where each of the $x_i(s)$ and $y_i(s)$ are polynomials in the parameter s. As the parameter s is varied a curve is traced out.

12.3. SNAKE PROPERTIES

From now on snakes will be referred to as the parametric curve $\underline{u}(s) = (x(s), y(s))$ where s is assumed to vary between 0 and 1. What properties should an 'edge hugging' snake have?

- The snake must be 'driven' by the image. That is, it must be able to detect an edge in the image and align itself with the edge. One way of achieving this is to try to position the snake such that the average 'edge strength' (however that may be measured) along the length of the snake is maximized. If the measure of edge strength is $F(x,y) \geq 0$ at the image point (x,y) then this amounts to saying that the snake $\mathbf{u}(s)$ is to be chosen in such a way that the functional

$$\int_{s=0}^{s=1} F(x(s), y(s)) ds \qquad 12.1$$

is maximized. This will ensure that the snake will tend to mould itself to edges in the image if it finds them, but does not guarantee that it will find them in the first place. Given an image the functional may have

---

[1] The use of the term 'dynamically' is strictly speaking incorrect. The way that snakes have been programmed in the past make them appear as if they are moving in time, but this is just a product of the implementation. Snakes are solutions to static problems; time does not enter into the formulation.

many local minima (a static problem); finding them is where the 'dynamics' arises. An edge detector applied to an image will tend to produce an edge map consisting of mainly thin edges. This means that the edge strength function tends to be zero at most places in the image, apart from on a few lines. As a consequence a snake placed some distance from an edge may not be attracted towards the edge because the edge strength is effectively zero at the snake's initial position. To help the snake come under the influence of an edge the edge image is blurred to broaden the width of the edges.

- If an elastic band were held around a convex object and then let go, the band would contract until the object prevented it from doing so further. At this point the band would be moulded to the object, thus describing the boundary. Two forces are at work here — firstly that providing the natural tendency of the band to contract, and secondly the opposing force provided by the object. The band contracts because it tries to minimize its elastic energy due to stretching. If the band is described by the parametric curve $\mathbf{u}(s) = (x(s), y(s))$ then the elastic energy at any point $\hat{s}$ is proportional to

$$\left(\frac{d\mathbf{u}}{ds}\bigg|_{\hat{s}}\right)^2 = \left(\frac{dx}{ds}\bigg|_{\hat{s}}\right)^2 + \left(\frac{dy}{ds}\bigg|_{\hat{s}}\right)^2$$

That is, the energy is proportional to the square of how much the curve is being stretched at that point. The elastic band will take up a configuration so that the elastic energy along its entire length, given the constraint of the object, is minimized. Hence the elastic band assumes the shape of the curve $\mathbf{u}(s) = (x(s), y(s))$ where $\mathbf{u}(s)$ minimizes the functional

$$\int_{s=0}^{s=1} \left\{ \left(\frac{dx}{ds}\right)^2 + \left(\frac{dy}{ds}\right)^2 \right\} ds \qquad 12.2$$

subject to the constraints of the object. Ideally, closed snakes should have analogous behaviour: that is, to have a tendency to contract, but to be prevented from doing so by the objects in an image. To model this behaviour the parametric curve for the snake is chosen so that the functional, equation (12.2), tends to be minimized. If in addition the forcing term, equation (12.1), were included then the snake would be prevented from contracting 'through objects' as it would be attracted toward their edges. The attractive force would also tend to pull the snake into the hollows of a concave boundary, provided that the restoring 'elastic force' was not too great.

- One of the properties of edges that is difficult to model is their behaviour when they can no longer be seen. If someone were looking at a car and a person stood in front of it, few people would have any difficulty imagining the contours of the edge of the car that were occluded. They would be 'smooth' extensions of the contours either side of the person. If the above elastic band approach were adopted it would be found that the band formed a straight line where the car was occluded (because it tries to minimize energy, and thus length, in this situation). If however the band had some stiffness (that is a resistance to bending, as for example displayed by a flexible bar) then it would tend to form a smooth curve in the occluded region of the image and be tangential to the boundaries on either side (Fig. 12.1).

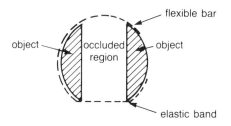

Fig. 12.1 Extreme extensions to occluded objects.

Again a flexible bar tends to form a shape so that its elastic energy is minimized. The elastic energy in bending is dependent on the curvature of the bar, that is the second derivatives. To help force the snake to emulate this type of behaviour the parametric curve $\mathbf{u}(s) = (x(s), y(s))$ is chosen so that it tends to minimize the functional

$$\int_{s=0}^{s=1} \left\{ \left(\frac{d^2x}{ds^2} \right)^2 + \left(\frac{d^2y}{ds^2} \right)^2 \right\} ds \qquad 12.3$$

which represents a psuedo-bending energy term. Of course, if a snake were made too stiff then it would be difficult to force it to conform to highly curved boundaries under the action of the forcing term equation (12.1).

Three desirable properties of snakes have now been identified. To incorporate all three into the snake at once the parametric curve $\mathbf{u}(s) = (x(s), y(s))$ representing the snake is chosen so that it minimizes the functional

$$I(x(s),y(s)) = \int_{s=0}^{s=1} \{\alpha(s)\{\left(\frac{d^2x}{ds^2}\right)^2 + \left(\frac{d^2y}{ds^2}\right)^2\}$$

$$+ \beta(s)\{\left(\frac{dx}{ds}\right)^2 + \left(\frac{dy}{ds}\right)^2\}$$

$$- F(x(s),y(s))\}ds \qquad 12.4$$

Here the terms $\alpha(s) > 0$ and $\beta(s) \geq 0$ represent respectively the amount of stiffness and elasticity that the snake is to have. It is clear that if the snake approach is to be successful then the correct balance of these parameters is crucial. Too much stiffness and the snake will not correctly hug the boundaries; too much elasticity and closed snakes will be pulled across boundaries and contract to a point, or may even break away from boundaries at concave regions. The negative sign in front of the forcing term is because minimizing $-\int F(x,y)ds$ is equivalent to maximizing $\int F(x,y)ds$.

As it stands, minimizing the functional, equation (12.4), is trivial. If the snake is not closed then the solution degenerates into a single point $(x(s),y(s))$ = constant where the point is chosen to minimize the edge strength $F(x(s),y(s))$. Physically, this is because the snake will tend to pull its two end points together in order to minimize the elastic energy, and thus shrink to a single point. The global minimum is attained at the point in the image where the edge strength is largest[2]. To prevent this from occurring it is necessary to fix the positions of the ends of the snake in some way; i.e. 'boundary conditions' are required. It turns out to be necessary to fix more than just the location of the end points and two further conditions are required for a well-posed problem. A convenient condition is to impose zero curvature at each end point.

Similarly, the global minimum for a closed-loop snake occurs when it contracts to a single point. However, in contrast to an open-loop snake, additional boundary conditions cannot be applied to eliminate the degenerate solution. The degenerate solution is the true global minimum.

---

[2] By a global minimum is meant a curve $u(s)$ such that $I(u(s)) \leq I(U(s))$ for all $U(s)$. Such a curve may not be unique (as the edge strength can take on its maximum value at one or more points) and so strictly speaking does not represent a true global minimum.

12.4. FORMULATION

Clearly the ideal situation is to seek a local minimum in the locality of the initial position of the snake. In practice the problem that is solved is weaker than this: find a curve $\hat{\mathbf{u}}(s) = (\hat{x}(s), \hat{y}(s)) \in H^2[0,1] \times H^2[0,1]$ such that

$$\frac{\partial I(\hat{\mathbf{u}}(s) + \epsilon \mathbf{v}(s))}{\partial \epsilon}\bigg|_{\epsilon=0} = 0 \quad \forall \mathbf{v}(s) \in H_0^2[0,1] \times H_0^2[0,1] \quad \quad 12.5$$

Here $H^2[0,1]$ denotes the class of functions defined on [0,1] that have 'finite energy' in the second derivatives (that is the integral of the square of second derivatives exists [4]) and $H_0^2[0,1]$ is the class of real valued functions in $H^2[0,1]$ that are zero at $s=0$ and $s=1$. To see how this relates to finding a local minimum consider $\hat{\mathbf{u}}(s)$ to be a local minimum and $\hat{\mathbf{u}}(s) + \epsilon \mathbf{v}(s)$ to be a perturbation about the minimum that satisfies the same boundary conditions (because $\mathbf{v}(0) = \mathbf{v}(1) = 0$). Clearly, considered as a function of ϵ, $I(\epsilon) = I(\hat{\mathbf{u}}(s) + \epsilon \mathbf{v}(s))$ is a minimum at $\epsilon = 0$. Hence the derivate of $I(\epsilon)$ must be zero at $\epsilon = 0$. Equation (12.5) is therefore a *necessary* condition for a local minimum. Although solutions to equation (12.5) are not guaranteed to be minima (see Appendix) it has been found in practice that solutions are indeed minima.

It is easily shown (see Appendix) that equation (12.5) is equivalent to two other problems, each of which is simpler to solve.

Problem 1

Find a curve $(\hat{x}(s), \hat{y}(s)) \in H^2[0,1] \times H^2[0,1]$ such that

$$\int_{s=0}^{s=1} \left\{ \alpha(s) \frac{d^2\hat{x}}{ds^2} \frac{d^2\eta}{ds^2} + \beta(s) \frac{d\hat{x}}{ds} \frac{d\eta}{ds} - \frac{1}{2} \eta \frac{\partial F}{\partial x}\bigg|_{(\hat{x},\hat{y})} \right\} ds = 0$$

$$\forall \eta(s) \in H^2[0,1] \quad \quad 12.6$$

and

$$\int_{s=0}^{s=1} \left\{ \alpha(s) \frac{d^2\hat{y}}{ds^2} \frac{d^2\zeta}{ds^2} + \beta(s) \frac{d\hat{y}}{ds} \frac{d\zeta}{ds} - \frac{1}{2} \zeta \frac{\partial F}{\partial y}\bigg|_{(\hat{x},\hat{y})} \right\} ds = 0$$

$$\forall \zeta(s) \in H^2[0,1] \quad \quad 12.7$$

$\hat{x}(0), \hat{y}(0), \hat{x}(1), \hat{y}(1)$ given.

Problem 2

Find a curve $(\hat{x}(s), \hat{y}(s)) \in C^4[0,1] \times C^4[0,1]$ that satisfies the pair of fourth-order ordinary differential equations

$$-\frac{d^2}{ds^2}\left\{\alpha(s)\frac{d^2\hat{x}}{ds^2}\right\} + \frac{d}{ds}\left\{\beta(s)\frac{d\hat{x}}{ds}\right\} + \frac{1}{2}\frac{\partial F}{\partial x}\bigg|_{(\hat{x},\hat{y})} = 0 \qquad 12.8$$

$$-\frac{d^2}{ds^2}\left\{\alpha(s)\frac{d^2\hat{y}}{ds^2}\right\} + \frac{d}{ds}\left\{\beta(s)\frac{d\hat{y}}{ds}\right\} + \frac{1}{2}\frac{\partial F}{\partial y}\bigg|_{(\hat{x},\hat{y})} = 0 \qquad 12.9$$

together with the boundary conditions $\hat{x}(0), \hat{y}(0), \hat{x}(1), \hat{y}(1)$ given, and

$$\frac{d^2\hat{x}}{ds^2}\bigg|_{s=0} = \frac{d^2\hat{y}}{ds^2}\bigg|_{s=0} = \frac{d^2\hat{x}}{ds^2}\bigg|_{s=1} = \frac{d^2\hat{y}}{ds^2}\bigg|_{s=1} = 0 \qquad 12.10$$

Here the statement of each problem adopts boundary conditions to cover the case of a fixed-end snake. However, if the snake is to form a closed loop the above boundary conditions are replaced by periodicity conditions. The first problem is known as a 'variational problem' and is a 'weak form' [5] of the differential equations (12.6) and (12.7). The most natural way to tackle problem 1 is to use a finite element approach [5]. Problem 2 can easily be solved by using a different but related technique: finite differences. A description of each approach for this particular problem is given in the following two sections.

12.5. FINITE ELEMENTS

In the statement of problem 2, two sets of curves occur: the 'trial' curves $(x(s), y(s))$ and the test curves $(\eta(s), \zeta(s))$. The trial curves are said to belong to the trial space and the test curves to the test space. For the current problem both spaces are chosen to be the same, namely $H^2[0,1] \times H^2[0,1]$. The trial and test curves get their names from the fact that notionally a trial solution can be substituted into equations (12.6) and (12.7) and tested to see if it satisfies the equations by cycling through all the test functions. In practice this cannot be done as the spaces are infinite-dimensional — an alternative approximate approach is therefore sought.

As the trial and test spaces consist of relatively smooth functions (twice differentiable in the weak sense) it is possible to approximate them closely using piecewise polynomials of low degree. Thus, instead of using the full trial and test spaces a finite-dimensional subspace consisting of piecewise polynomials of a given degree is used. This space is called the 'approximating subspace' and the idea is to replace the true minimizing curve in the full space with the 'nearest' curve in the approximating subspace.

The approximate solution $(\hat{x}_h(s), \hat{y}_h(s))$ is expressed in terms of a sum of basis functions for the approximating subspace. The choice of basis functions (and hence choice of subspace) is governed by two main criteria — computational efficiency and admissibility. Computational efficiency is achieved by the use of a 'local' polynomial basis. That is, the 'overlap' between basis functions in the interval [0,1] is small, and it turns out that this ensures that calculating the contribution of each basis function in the expansion is relatively simple. Admissibility means that the basis functions must belong to the trial and test spaces, i.e. they must have finite energy in the second derivatives.

In terms of the basis function expansion the approximate solution is

$$\hat{x}_h(s) = \sum_{i=1}^{N} x_i \phi_i^{(1)}(s) + \sum_{i=1}^{N} x_i' \phi_i^{(2)}(s) \qquad 12.11$$

$$\hat{y}_h(s) = \sum_{i=1}^{N} y_i \phi_i^{(1)}(s) + \sum_{i=1}^{N} y_i' \phi_i^{(2)}(s) \qquad 12.12$$

where the coefficients x_i, x_i', y_i and y_i' in the sum are to be found. The basis functions are chosen as follows: divide the interval [0,1] into $N-1$ subintervals of 'elements'

$$0 = s_1 < s_2 < \ldots < s_{N-2} < s_{N-1} < s_N = 1$$

and at the ith 'node' define two functions as shown in Fig. 12.2.

Here the function $\phi_i^{(1)}$ is a cubic polynomial on each of the intervals $[s_{i-1}, s_i]$ and $[s_i, s_{i+1}]$, is zero outside the intervals, takes the value one at s_i, is zero at s_{i-1} and s_{i+1} and has zero derivatives at s_{i-1}, and s_{i+1}. The function $\phi_i^{(2)}$ is also a cubic polynomial on the same intervals, is zero outside the intervals, takes the value zero at s_{i-1}, s_i and s_{i+1}, has zero derivatives at s_{i-1} and s_{i+1} and has derivative one at s_i. These basis functions, known as 'Hermite cubics' [5], have continuous first derivatives and a finite number of jump discontinuities in the second derivative: hence they are admissible. It is clear from equations (12.11) and (12.12) that the unknown (x_i, y_i) represents points in the image that the snake interpolates and (x_i', y_i') represent the first derivative of the snake at those points.

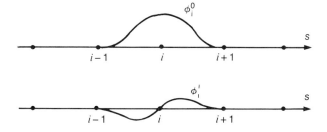

Fig. 12.2 Hermite basis functions.

In the case of the fixed end point snake x_1, y_1, x_N and y_N are known in advance through the boundary conditions. It turns out however that it is not necessary explicitly to impose the zero curvature boundary conditions as this is naturally enforced by the weak formulation of equations (12.6) and (12.7). For the closed-loop snake periodic boundary conditions are applied by introducing 'fictitious' basis functions $\phi_0^{(1)}$, $\phi_0^{(2)}$, $\phi_{N+1}^{(1)}$ and $\phi_{N+1}^{(2)}$ and assigning

$$x_0 = x_N, \quad x'_0 = x'_N, \quad x_{N+1} = x_1, \quad x'_{N+1} = x'_1$$

$$y_0 = y_N, \quad y'_0 = y'_N, \quad y_{N+1} = y_1, \quad y'_{N+1} = y'_1$$

If the basis function expansion is substituted into the weak forms, equations (12.6) and (12.7), and the test functions are taken to be each basis function in turn, the result is a set of algebraic equations for the unknowns x_i, x'_i, y_i and y'_i. (This approach is known as Galerkin's method [5].)

$$K\mathbf{x} = \mathbf{f}(\mathbf{x},\mathbf{y}) \text{ and } K\mathbf{y} = \mathbf{g}(\mathbf{x},\mathbf{y}) \qquad 12.13$$

The matrix K, known as the stiffness matrix, is blocked into 2×2 sub-matrices $K_{ij} = \{k_{pq}^{ij}\}$ where

$$k_{pq}^{ij} = \int_{s=0}^{s=1} \left\{ \alpha(s) \frac{d^2\phi_i^{(p)}}{ds^2} \frac{d^2\phi_j^{(q)}}{ds^2} + \beta(s) \frac{d\phi_i^{(p)}}{ds} \frac{d\phi_j^{(q)}}{ds} \right\} ds$$

for $i, j = 1,\ldots,N \quad p,q = 1,2$ \qquad 12.14

Because the basis functions do not overlap it follows that most of the sub-matrices turn out to be zero. Furthermore, $K_{ji} = K_{ij}^T$ and hence the matrix is symmetric.

Because of the different types of boundary condition the stiffness matrix takes on a slightly different form depending on whether the snake has fixed ends or forms a loop. For a closed-loop snake the matrix is of the form

$$K = \begin{Bmatrix} K_{11} & K_{12} & & & & & K_{1N} \\ K_{21} & K_{22} & K_{23} & & & & \\ & K_{32} & K_{33} & K_{34} & & & \\ & & \cdot & \cdot & \cdot & & \\ & & & \cdot & \cdot & \cdot & \\ K_{N1} & & & & & K_{NN-1} & K_{NN} \end{Bmatrix}$$

and for a fixed-end snake is

$$K = \begin{Bmatrix} K_{11}^L & K_{12}^L & & & & \\ K_{21} & K_{22} & K_{23} & & & \\ & K_{32} & K_{33} & K_{34} & & \\ & & \cdot & \cdot & \cdot & \\ & & & \cdot & \cdot & \cdot \\ & & & & K_{NN-1}^L & K_{NN}^L \end{Bmatrix}$$

where K_{11}^L, K_{12}^L, K_{NN-1}^L and K_{NN}^L are 1×2 matrices consisting of the lower row of the matrices K_{11}, K_{12}, K_{NN-1} and K_{NN} respectively. Similarly the right-hand vectors **f** and **g** are composed of two-dimensional sub-vectors $\mathbf{f}_i = \{f_p^i\}$ and $\mathbf{g}_i = \{g_p^i\}$ where

$$f_p^i = -\frac{1}{2} \int_{s=0}^{s=1} \phi_i^p \frac{\partial F}{\partial x}\bigg|_{(\hat{x}_h, \hat{y}_h)} ds \quad \text{for } i = 1,\ldots,N \; p = 1,2$$

12.15

$$g_p^i = -\frac{1}{2} \int_{s=0}^{s=1} \phi_i^p \frac{\partial F}{\partial y}\bigg|_{(\hat{x}_h, \hat{y}_h)} ds \quad \text{for } i = 1,\ldots,N \; p = 1,2$$

12.16

For a closed-loop snake the right-hand vectors are of the form

$$\mathbf{f} = (\mathbf{f}_1, \mathbf{f}_2, \ldots \mathbf{f}_N)^T$$

and similarly for **g**. For a fixed-end snake the right-hand vector is

$$\mathbf{f} = (f_1^1 - k_{21}^{11} x_1, \mathbf{f}_2, \ldots \mathbf{f}_{N-1}, f_1^N - k_{12}^{NN} x_N)^T$$

and similarly for **g**. The first and last elements in the vector arise through the application of the fixed-end boundary conditions. For elements of unequal size or $\alpha(s)$ and $\beta(s)$ variable the integrals in equation (12.14) would normally be calculated numerically [5]. However, for uniform elements of length h and α and β constant they are easily evaluated analytically. In this case the submatrices turn out to be

$$K_{11} = \frac{\alpha}{h^3} \begin{pmatrix} 12 & 6h \\ 6h & 4h^2 \end{pmatrix} + \frac{\beta}{30h} \begin{pmatrix} 36 & 3h \\ 3h & 4h^2 \end{pmatrix}$$

$$K_{12} = \frac{\alpha}{h^3} \begin{pmatrix} -12 & 6h \\ -6h & 2h^2 \end{pmatrix} + \frac{\beta}{30h} \begin{pmatrix} -36 & 3h \\ -3h & -h^2 \end{pmatrix}$$

$$K_{22} = K_{33} = \ldots = K_{N-1\,N-1} = 2K_{11}$$

$$K_{NN} = K_{11}$$

$$K_{23} = K_{34} = \ldots = K_{N-1\,N} = K_{12}$$

$$K_{1N} = K_{12}^T$$

The unknowns in the system of equation (13) are

$$\mathbf{x} = (x_1, x'_1, x_2, x'_2, \ldots, x_N, x'_N)^T$$

$$\mathbf{y} = (y_1, y'_1, y_2, y'_2, \ldots, y_N, y'_N)^T$$

12.6. FINITE DIFFERENCES

The finite difference approach starts by discretizing the interval $[0,1]$ into $N-1$ equispaced subintervals of length $h = 1/(N-1)$ and defines a set of nodes $\{s_i\}_{i=1}^{i=N}$ where $s_i = (i-1)h$. The method seeks a set of approximations $\{(x_i, y_i)\}_{i=1}^{i=N}$ to $\{(x(s_i), y(s_i))\}_{i=1}^{i=N}$ by replacing the differential equations (12.8) and (12.9) in the continuous variables with a set of difference equations in the discrete variables [9]. Replacing the derivatives in equation (12.8) by difference approximations at the point s_i gives

$$-\frac{1}{h^2}\left\{\alpha_{i+1}\frac{(x_{i+2}-2x_{i+1}+x_i)}{h^2} - 2\alpha_i\frac{(x_{i+1}-2x_i+x_{i-1})}{h^2}\right.$$
$$\left. + \alpha_{i-1}\frac{(x_i-2x_{i-1}+x_{i-2})}{h^2}\right\}$$
$$+\frac{1}{h}\left\{\beta_{i+1}\frac{(x_{i+1}-x_i)}{h} - \beta_i\frac{(x_i-x_{i-1})}{h}\right\}$$
$$+\frac{1}{2}\frac{\partial F}{\partial x}\bigg|_{(x_i,y_i)} = 0 \qquad \text{for } i=3,4,\ldots,N-2 \qquad 12.17$$

where $\alpha_i = \alpha(s_i)$ and $\beta_i = \beta(s_i)$. Note that the difference equation only holds at internal nodes in the interval where the indices referenced lie in the range 1 to N. Collecting like terms together, equation (12.17) can be written as

$$a_i x_{i-2} + b_i x_{i-1} + c_i x_i + d_i x_{i+1} + e_i x_{i+2} = f_i$$

where

$$a_i = -\frac{\alpha_{i-1}}{h^4}$$

$$b_i = \frac{2\alpha_i}{h^4} + \frac{2\alpha_{i-1}}{h^4} + \frac{\beta_i}{h^2}$$

$$c_i = -\left\{\frac{\alpha_{i+1}}{h^4} + \frac{4\alpha_i}{h^4} + \frac{\alpha_{i-1}}{h^4} + \frac{\beta_{i+1}}{h^2} + \frac{\beta_i}{h^2}\right\}$$

$$d_i = \frac{2\alpha_{i+1}}{h^4} + \frac{2\alpha_i}{h^4} + \frac{\beta_{i+1}}{h^2}$$

$$e_i = -\frac{\alpha_{i+1}}{h^4}$$

$$f_i = -\frac{1}{2}\frac{\partial F}{\partial x}\bigg|_{(x_i,y_i)}$$

Taking boundary conditions into account the finite difference approximations satisfy the following system of algebraic equations

$$K\mathbf{x} = \mathbf{f}(\mathbf{x},\mathbf{y}), \quad K\mathbf{y} = \mathbf{g}(\mathbf{x},\mathbf{y}) \qquad 12.18$$

As in the finite element approach, the structure of the matrices K and the right-hand vectors $\mathbf{f}$ and $\mathbf{g}$ are different depending on whether closed or open snake boundary conditions are used. If the snake is closed then fictitious nodes at s_0, s_{-1}, s_{N+1} and s_{N+2} are introduced and the difference equation (12.17) is applied at nodes 0, 1, $N-1$ and N. Periodicity implies that $x_0 = x_N$, $x_{-1} = x_{N-1}$, and $x_{N+1} = x_1$, $x_2 = x_{N+2}$. With these conditions in force the coefficient matrix becomes

$$K = \begin{Bmatrix} c_1 & d_1 & e_1 & & & & & a_1 & b_1 \\ b_2 & c_2 & d_2 & e_2 & & & & & a_2 \\ a_3 & b_3 & c_3 & d_3 & e_3 & & & & \\ & a_4 & b_4 & c_4 & d_4 & e_4 & & & \\ & & \cdot & \cdot & \cdot & \cdot & \cdot & & \\ & & & \cdot & \cdot & \cdot & \cdot & \cdot & \\ e_{N-1} & & & & & a_{N-1} & b_{N-1} & c_{N-1} & d_{N-1} \\ d_N & e_N & & & & & a_N & b_N & c_N \end{Bmatrix}$$

and the right-hand side vector is

$$(f_1, f_2, \ldots, f_N)^T$$

For fixed-end snakes fictitious nodes at s_0 and s_{N+1} are introduced and the difference equation (12.17) is applied at nodes s_1 and s_{N+1}. Two extra difference equations are introduced to approximate the zero curvature boundary conditions

$$\left.\frac{d^2x}{ds^2}\right|_{s_1} = \left.\frac{d^2x}{ds^2}\right|_{s_N} = 0$$

namely $x_0 - 2x_1 + x_2 = 0$ and $x_{N-1} - 2x_N + x_{N+1} = 0$. The coefficient matrix is now

$$K = \begin{Bmatrix} c_2-a_2 & d_2 & e_2 \\ b_3 & c_3 & d_3 & e_3 \\ a_4 & b_4 & c_4 & d_4 & e_4 \\ & a_5 & b_5 & c_5 & d_5 & e_5 \\ & & \cdot & \cdot & \cdot & \cdot \\ & & & \cdot & \cdot & \cdot & \cdot \\ & & & & a_{N-1} & b_{N-2} & c_{N-2} & d_{N-2} \\ & & & & & a_{N-1} & b_{N-1} & c_{N-1}-e_{N-1} \end{Bmatrix}$$

and the right-hand side vector is

$$(f_2 - (2a_2 + b_2)x_1, \quad f_3 - a_3 x_1, \quad f_4, \ldots$$

$$\ldots, f_{N-3}, \quad f_{N-2} - e_{N-2}x_N, \quad f_{N-1} - (2e_{N-1} + d_{N-1})x_N)^T$$

Similarly, a set of difference equations can be derived for equation (12.9).

12.7. SYSTEM SOLUTION

Whether finite elements or finite differences are used the end result is a system of nonlinear equations that need to be solved. In the case of finite elements the system comprises equation (12.13) and in the case of finite differences equation (12.18). With both methods the coefficient matrix has a similar structure (it is symmetric and positive definite, banded for the fixed-end snake, and for a closed-loop snake with periodic boundary conditions is banded, apart from a few off-diagonal entries) and the same method can be used in either case. As the system is nonlinear it is solved iteratively. The iteration performed is

$$\frac{(x_{n+1} - x_n)}{\gamma} + K x_{n+1} = f(x_n, y_n) \text{ for } n = 0, 1, 2, \ldots$$

$$\frac{(y_{n+1} - y_n)}{\gamma} + K y_{n+1} = g(x_n, y_n) \text{ for } n = 0, 1, 2, \ldots$$

12.19

where $\gamma > 0$ is a stabilization parameter. This can be rewritten as

$$(K + \frac{1}{\gamma}I)\mathbf{x}_{n+1} = \frac{1}{\gamma}\mathbf{x}_n + f(\mathbf{x}_n, \mathbf{y}_n) \text{ for } n = 0,1,2,\ldots$$

$$(K + \frac{1}{\gamma}I)\mathbf{y}_{n+1} = \frac{1}{\gamma}\mathbf{y}_n + g(\mathbf{x}_n, \mathbf{y}_n) \text{ for } n = 0,1,2,\ldots$$

This system has to be solved for each n. For a closed-loop snake the matrix on the left-hand side is difficult to invert directly because of the terms that are outside the main diagonal band destroy the band structure. In general the coefficient matrix K can be split into the sum of a banded matrix B plus a non-banded matrix A, $K = A + B$. For a fixed-end snake the matrix A would be zero. The system of equations is now solved for each n by performing the iteration

$$(B + \frac{1}{\gamma}I)\mathbf{x}_{n+1}^{(k+1)} = -A\mathbf{x}_{n+1}^{(k)} + \frac{1}{\gamma}\mathbf{x}_n + \mathbf{f}(\mathbf{x}_n, \mathbf{y}_n)$$

for $k = 0,1,2,\ldots$

$$(B + \frac{1}{\gamma}I)\mathbf{y}_{n+1}^{(k+1)} = -A\mathbf{y}_{n+1}^{(k)} + \frac{1}{\gamma}\mathbf{y}_n + \mathbf{g}(\mathbf{x}_n, \mathbf{y}_n)$$

for $k = 0,1,2,\ldots$

The matrix $(B + \frac{1}{\gamma}I)$ is a band matrix and can be expressed as a product of Cholesky factors LL^T [11]. The systems are solved at each stage by first solving

$$L\tilde{\mathbf{x}}_{n+1}^{(k+1)} = -A\mathbf{x}_{n+1}^{(k)} + \frac{1}{\gamma}\mathbf{x}_n + \mathbf{f}(\mathbf{x}_n, \mathbf{y}_n)$$

$$L\tilde{\mathbf{y}}_{n+1}^{(k+1)} = -A\mathbf{y}_{n+1}^{(k)} + \frac{1}{\gamma}\mathbf{y}_n + \mathbf{g}(\mathbf{x}_n, \mathbf{y}_n)$$

followed by

$$L^T\mathbf{x}_{n+1}^{(k+1)} = -\tilde{\mathbf{x}}_{n+1}^{(k+1)}$$

$$L^T\mathbf{y}_{n+1}^{(k+1)} = -\tilde{\mathbf{y}}_{n+1}^{(k+1)}$$

Notice that the Cholesky decomposition only has to be performed once.

260 HEAD BOUNDARY LOCATION USING SNAKES

12.8. RESULTS

In this section some results are presented from the implementation of a closed-loop snake to find the boundary of a head in a set of 28 head-and-shoulders images. Details of the choice of the edge strength function are given in a companion chapter [2].

The initial position for the snake (Fig. 12.3) in the iteration, equation (12.19), was taken to be the extreme edges of the image with the nodes equispaced and the corners slightly rounded. As the iteration proceeds successive iterates represent a sequence with the snake contracting due to the predominance of the elastic force (Figs. 12.4 and 12.5). The snake finally reaches an equilibrium position when it is aligned with edges in the image (Fig. 12.6). In 15 of the images the equilibrium position resulted in the snake hugging the boundary of the head. In nine of the images the snake located most of the head boundary but became trapped on features external to the head (for example a shirt collar, Fig. 12.7). In the remaining four cases the snake penetrated the head boundary and contracted on to features interior to the face. In all cases the same elastic and bending parameters were used and the results clearly demonstrate the need to select the parameters appropriately. Some methods for detecting when the snake has become stuck are outlined in chapter 11 and work is currently in progress to control the snake parameters adaptively to unstick it.

Fig. 12.3.

RESULTS 261

Fig. 12.4.

Fig. 12.5.

Fig. 12.6.

Fig. 12.7.

APPENDIX

Problem solving

Using the techniques of the calculus of variations [10] it is easily demonstrated that solving equation (12.5) is equivalent to solving either problem 1 or problem 2 of section 12.4. Let $\hat{u}(s)$ be a stationary curve of the functional $I(\mathbf{u}(s))$ and let $\hat{\mathbf{u}}(s) + \epsilon \mathbf{v}(s) = (\hat{x}(s) + \epsilon \eta(s), \hat{y}(s) + \epsilon \zeta(s))$ be an admissible perturbation about the stationary curve (i.e.

$\eta(0) = \zeta(0) = \eta(1) = \zeta(1) = 0$, so that the position of the perturbed curve agrees with the minimizing curve at the end points). Substituting the expression for the perturbed curve into equation (12.4) gives

$I(\hat{x}(s) + \epsilon\eta(s), \hat{y}(s) + \epsilon\zeta(s)) =$

$$\int_{s=0}^{s=1} \left\{ \alpha(s) \left\{ \left(\frac{d^2\hat{x}}{ds^2}\right)^2 + 2\epsilon \frac{d^2\hat{x}}{ds^2}\frac{d^2\eta}{ds^2} + \epsilon^2\left(\frac{d^2\eta}{ds^2}\right)^2 \right.\right.$$

$$\left. + \left(\frac{d^2\hat{y}}{ds^2}\right)^2 + 2\epsilon \frac{d^2\hat{y}}{ds^2}\frac{d^2\zeta}{ds^2} + \epsilon^2\left(\frac{d^2\zeta}{ds^2}\right)^2 \right\}$$

$$+ \beta(s)\left\{ \left(\frac{d^2\hat{x}}{ds^2}\right)^2 + 2\epsilon \frac{d\hat{x}}{ds}\frac{d\eta}{ds} + \epsilon^2\left(\frac{d\eta}{ds}\right)^2 \right.$$

$$\left. + \left(\frac{d\hat{x}}{ds}\right)^2 + 2\epsilon \frac{d\hat{y}}{ds}\frac{d\zeta}{ds} + \epsilon^2\left(\frac{d\zeta}{ds}\right)^2 \right\}$$

$$\left. - F(x(s) + \epsilon\eta(s), y(s) + \epsilon\zeta(s)) \right\} ds \qquad 12.\text{A1}$$

Expanding the final term as a Taylor series

$F(\hat{x} + \epsilon\eta, \hat{y} + \epsilon\zeta) =$

$$F(\hat{x},\hat{y}) + \left.\frac{\partial F}{\partial x}\right|_{(\hat{x},\hat{y})} \epsilon\eta + \left.\frac{\partial F}{\partial y}\right|_{(\hat{x},\hat{y})} \epsilon\zeta + O(\epsilon^2)$$

and collecting like powers of ϵ in equation (12.A1) yields

$I(\hat{x}(s) + \epsilon\eta(s), \hat{y}(s) + \epsilon\zeta(s)) = I(\hat{x}(x), \hat{y}(s))$

$$+ 2\epsilon \int_{s=0}^{s=1} \left\{ \alpha(s) \frac{d^2\hat{x}}{ds^2}\frac{d^2\eta}{ds^2} + \beta(s) \frac{d\hat{x}}{ds}\frac{d\eta}{ds} - \frac{1}{2}\eta \left.\frac{\partial F}{\partial x}\right|_{(\hat{x},\hat{y})} \right\} ds$$

$$+ 2\epsilon \int_{s=0}^{s=1} \left\{ \alpha(s) \frac{d^2\hat{y}}{ds^2}\frac{d^2\zeta}{ds^2} + \beta(s) \frac{d\hat{y}}{ds}\frac{d\zeta}{ds} - \frac{1}{2}\zeta \left.\frac{\partial F}{\partial y}\right|_{(\hat{x},\hat{y})} \right\} ds$$

$+ O(\epsilon^2)$ \hfill 12.A2

From the definition of the stationary value it directly follows that the two integrals in equation (12.A2) are identically zero. That is, equations (12.6) and (12.7) of problem 1 hold. Performing integration by parts twice on equation (12.6) gives

$$\left[\alpha(s)\frac{d\hat{x}}{ds}\eta(s)\right]_{s=0}^{s=1} - \int_{s=0}^{s=1}\frac{d}{ds}\{\alpha(s)\frac{d\hat{x}}{ds}\}\eta(s)ds$$

$$+ \left[\beta(s)\frac{d^2\hat{x}}{ds^2}\frac{d\eta}{ds}\right]_{s=0}^{s=1} - \left[\frac{d}{ds}\{\beta(s)\frac{d^2\hat{x}}{ds^2}\}\eta(s)\right]_{s=0}^{s=1}$$

$$+ \int_{s=0}^{s=1}\frac{d^2}{ds^2}\{\beta(s)\frac{d^2\hat{x}}{ds^2}\}\eta(s)ds - \frac{1}{2}\int_{s=0}^{s=1}\frac{\partial F}{\partial x}\bigg|_{(\hat{x},\hat{y})}\eta(s)ds$$

$$= 0 \quad \forall \eta \in H^2[0,1] \qquad 12.A3$$

From the boundary conditions the first, third and fourth terms of equation (12.A3) are zero and so it follows that

$$\int_{s=0}^{s=1}\{\frac{d^2}{ds^2}\{\alpha(s)\frac{d^2\hat{x}}{ds^2}\}\eta(s) - \frac{d}{ds}\{\beta(s)\}\eta(s)$$

$$- \frac{1}{2}\frac{\partial F}{\partial x}\bigg|_{(\hat{x},\hat{y})}\eta(s)\}ds = 0 \qquad \forall \eta \in H^2[0,1] \qquad 12.A4$$

Because equation (12.A4) holds for all $\eta(s)$ it follows that equation (12.8) holds. An identical analysis shows that equation (12.9) also holds. In a similar fashion, starting with equations (12.8) and (12.9), multiplying through by test functions and integrating, equations (12.5), (12.6) and (12.7) may be deduced.

To guarantee that the stationary point is a minimum it must be shown that the second variation (that is, the coefficient of ϵ^2 in the expansion) is always positive. For a particular choice of edge strength function this may be possible. However, it can be noted that as the contribution to the second variation from the elastic and bending terms is always non-negative then a negative influence of the forcing term may be partially offset. This may explain why only local minima were found in the experimental work.

REFERENCES

1. Pratt W K: 'Digital image processing', Wiley-Interscience (1978).
2. Welsh W J, Searby S, Waite J B: 'Model based image coding', Br Telecom Technol J, 8, No 3 (July 1990).

3. Kass M, Witkin A, Terpozopoulus D: 'Snakes: active contour models', International Journal of Computer Vision, 321-331 (1988).

4. Oden J T, Carey G F: 'Finite elements — mathematical aspects', Prentice Hall (1983).

5. Strang G, Fix G J: 'An analysis of the finite element method', Prentice Hall (1973).

6. Becker E B, Carey G F, Oden J T: 'Finite elements — an introduction', Prentice Hall (1981).

7. Carey G F, Oden J T: 'Finite elements — a second course', Prentice Hall (1983).

8. Carey G F, Oden J T: 'Finite elements — computational aspects', Prentice Hall (1984).

9. Keller H B: 'Numerical methods for two-point boundary value problems', Blaisdell (1968).

10. Cragg J W: 'Calculus of variations', Allen & Unwin Ltd (1973).

11. Johnson L W, Riess R D: 'Numerical analysis', Addison-Wesley, second edition (1982).

13

IMAGE DATA COMPRESSION USING FRACTAL TECHNIQUES

J M Beaumont

13.1. INTRODUCTION

The word 'fractal' is derived from the phrase 'fractionally dimensional'. Examples of entities with fractional dimensions are coastlines and mountain ranges. Given a map of Britain and a pair of dividers set to 2 cm one can get a measure of the coastline's length. However, if the dividers are set to 1 cm the coast-line appears to have increased in length. As the coastline is inspected at greater magnifications more complexity emerges. The fractal dimension of a coast-line is a measure of its complexity and lies in the range $1-2$ D.

If Coastline__length varies as $(\text{Divider\_\_length})^{(1-FD)}$ then FD is the fractal dimension of the coastline.

An example of a curve with a fractal dimension of 2 is the Peano curve shown in Fig. 13.1. Inspecting the curve at higher and higher magnifications reveals greater levels of complexity — the detail eventually completely fills the square. In fact the curve has infinite complexity, a property shared by all fractals. As the square has dimension 2, and the Peano curve completely fills the square, it is reassuring that the Peano curve has dimension 2 also. Peano curves can be designed that fill out a cube and these will have a fractal dimension of 3.

INTRODUCTION 267

Fig. 13.1 2-D Peano curve.

Fractal techniques have been used for some time to create amazingly realistic computer graphics, such as mountain ranges and tree foliage. They are based on algorithms that recursively or iteratively generate patterns based on a local rule (local in the sense of the point reached in the algorithm rather then spatially local) [4,5]. Quite simple rules can generate seemingly complicated patterns. In this sense the pattern is encoded into the local rule.

One of these algorithms is based on affine transforms. The general form of a 2-D affine transform is as follows:

$$X = Ax + By + E$$
$$Y = Cx + Dy + F$$

An affine transform is a change of co-ordinate system. Points in the x,y plane are related to points in the X,Y plane by the above pair of equations. This transform has the following properties:

- co-ordinate translation and scaling, because it contains the set of transforms given by:

$$X = Ax \quad\quad + E$$
$$Y = \quad\quad Dy + F$$

- co-ordinate rotation, because it allows the set of transforms given by:

$$X = x\cos\phi - y\sin\phi$$
$$Y = x\sin\phi + y\cos\phi$$

- five degrees of freedom have been accounted for so far; however an affine transform has six variables A,B,C,D,E,F. The last degree of freedom allows for skewing of the 2-D plane, e.g. turning a rectangle into a parallelogram:

$$X = x + By$$
$$Y = \quad y$$

Generally, an affine transform can be used to map all the points x,y in one parallelogram into all the points X,Y in another parallelogram. Parallel lines are mapped on to other parallel lines, but angles are not preserved.

Affine transforms can be used for creating fractal pictures. Figure 13.2 shows a fractal fern generated by a set of four affine transforms. This set is called an iterated function system (IFS) code for the fern [1]. Each of the affine transforms P, Q, R, and S maps the large square on to one of the bold parallelograms (S has zero width). These parallelograms recursively map on to four smaller ones until their size is reduced to that of a pixel display element, at which point a dot is plotted. The resulting image is that of the fern shown. This is called the deterministic algorithm.

From the above description, it has been implied that all parallelograms eventually become small enough to plot. This is ensured by only allowing affine transforms that are contractive: that is, the area of a parallelogram after applying the affine transform is less than before. This puts a constraint on the possible values of A, B, C, and D:

$\text{abs}(AD - BC) < 1$

A stochastic method can be used for rendering fractal images. This gives them a textured appearance, and is also faster than the deterministic algorithm. An algorithm is used where one of the four affine transforms is chosen at random. Parts of the fern are seemingly visited and plotted in a probabilistic manner with a 'flying spot'. In Fig. 13.2, imagine that at a given point in time the spot has plotted the point at the apex of the fern. The next point to be plotted depends on which affine transform is chosen. In fact the algorithm chooses at random. Let it choose R, which represents the bottom-left leaf. The spot will then plot a point at the apex of the bottom-left leaf. This corresponds to the mapping of the top-right corner of the large square to the top corner of parallelogram R. The spot is now approximately a quarter of the way up from the bottom of the square. If affine transform S were now chosen at random, the spot would move to plot a point a quarter of the way up the stem (collapsed parallelogram S).

Initially the spot can be anywhere within the large square, but after a few iterations it will have converged on to the fern. Consequently the general appearance of the picture for an IFS code is apparent at an early point, but as the algorithm progresses it becomes less and less efficient as it replots points. Also, unless a perfect random-number generator is used, some points never get visited at all, no matter how long the program runs. However, unlike the deterministic algorithm described above, the stochastic algorithm does not need to evaluate the size of a parallelogram at every iteration, and consequently is numerically less complex.

The picture of the fern in Fig. 13.2 is described by four affine transforms or 24 numbers in total. Those 24 numbers are effectively compressed encoded data representing the fern picture. Conventional compression techniques, such as discrete cosine transform [3], would result in a much less compact description. It is this magnitude of compression that has created interest in the image-proccessing community.

By extending affine transforms into three dimensions, to create fractal surfaces, the problem of compressing grey-scale images can be tackled. This fractal surface can be thought of as the grey level or brightness of a 2-D picture, and can be created by iteratively applying the following kind of transforms:

$$X = Ax + By + E$$
$$Y = Cx + Dy + F$$
$$Z = Gx + Hy + I$$

where the Z direction represents the grey level. The above set of equations are necessarily single-valued in Z, i.e. a pair of x,y values uniquely determines Z.

iterated function system code for fern

affine transformation
$$X = Ax + By + E$$
$$Y = Cx + Dy + F$$

		A	B	C	D	E	F
P	(main fern)	0.85	0.04	−0.04	0.85	0.02	0.08
Q	(right leaf)	−0.13	0.24	0.22	0.20	0.12	−0.27
R	(left leaf)	0.18	−0.24	0.21	0.20	−0.12	−0.30
S	(stalk)	0.0	0.0	0.0	0.16	0.0	−0.42

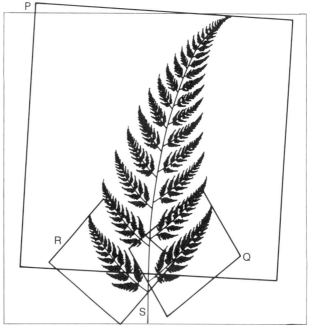

Fig. 13.2 Fractal fern.

The transform in Table 13.1 is for a grey level ramp with 'white' ($Z = 0.5$) at the top right corner, 'black' ($Z = -0.5$) at the bottom-left corner and the grey levels lying in between creating a line of 'mid-grey' ($Z = 0.0$) running from the top-left corner to the bottom-right corner.

The above example demonstrates that it is possible to produce a grey-level patch using three-dimensional affine transforms. A practical scheme for coding two-dimensional images will be described later, but first the properties of 2-D affine transforms will be explored.

13.2. REGION CODING

Figure 13.3 demonstrates the way in which affine transforms can be used to define a region; in this case the large triangle can be made to map into itself twice. The scaling and rotation used to map the large triangle into one of the smaller triangles defines the affine transformation. The affine transforms for the '3-4-5' triangle in Fig. 13.3 are shown in Table 13.2.

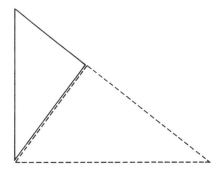

Fig. 13.3 Perfect collage of a 3-4-5 triangle.

In general any shape can be mapped into itself if small enough transforms are used. Usually though, a set of approximate mappings is good enough to reproduce a convincing representation of the original shape. Figure 13.4 shows a disc approximated rather crudely by four ellipses (transforms are allowed to overlap). The affine transforms used to generate the ellipses from the target disc are shown in Table 13.3.

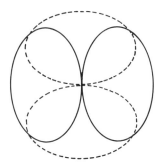

Fig. 13.4   Rough collage of a disc.

The property of being able to produce a more accurate representation of a region by using more accurate mappings is known as the collage theorem [1]. One would use nine ellipses to give a 'subjectively perfect disc', as indicated in Fig. 13.5. However, eight transforms are not necessarily required to draw a circle, as will be shown in section 13.3.

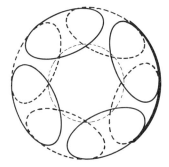

Fig. 13.5   Near-perfect collage of a disc.

13.3.   BOUNDARY CODING

By adaption of the deterministic algorithm (see section 13.1) boundaries rather than shapes can be encoded. This process will be explained by means of two examples, the circle and the straight line. Referring to Fig. 13.5, the affine transforms for the eight ellipses are shown in Table 13.4.

By using the deterministic algorithm with a modification, the outer circular boundary formed by the overlapping ellipses can be plotted.

BOUNDARY CODING

Table 13.1   IFS code for grey-level ramp

Transform	A	B	C	D	G	H	E	F	I
1 TOP RIGHT	0.5	0.0	0.0	0.5	0.25	0.25	0.25	0.25	0.25
2 TOP LEFT	0.5	0.0	0.0	0.5	0.25	0.25	−0.25	0.25	0.00
3 BOTTOM RIGHT	0.5	0.0	0.0	0.5	0.25	0.25	0.25	−0.25	0.00
4 BOTTOM LEFT	0.5	0.0	0.0	0.5	0.25	0.25	−0.25	−0.25	−0.25

Table 13.2   IFS code for 3-4-5 triangle

Transform	A	B	C	D	E	F
1 BOTTOM	0.64	−0.48	−0.48	−0.64	0.0	−0.375
2 LEFT	−0.36	−0.48	−0.48	0.36	−0.5	0.0

Table 13.3   IFS code for rough collage of disc

Transform	A	B	C	D	E	F
1 TOP	0.75	0.0	0.0	0.50	0.0	0.25
2 RIGHT	0.5	0.0	0.0	0.75	−0.25	0.0
3 BOTTOM	0.75	0.0	0.0	0.50	0.0	−0.25
4 LEFT	0.5	0.0	0.0	0.75	−0.25	0.0

Table 13.4   Transforms for eight ellipses forming circular boundary

Transform	A	B	C	D	E	F
1 RIGHT (UPPER)	0.240	−0.206	0.096	0.500	0.342	0.143
2 RIGHT (LOWER)	0.240	−0.206	−0.096	−0.500	0.342	−0.143
3 LEFT (UPPER)	−0.240	0.206	0.096	0.500	−0.342	0.143
4 LEFT (LOWER)	−0.240	0.206	−0.096	−0.500	−0.342	−0.143
5 TOP (RIGHT)	0.096	0.500	0.240	−0.206	0.143	0.342
6 TOP (LEFT)	−0.096	−0.500	0.240	−0.206	−0.143	0.342
7 BOTTOM (RIGHT)	0.096	0.500	−0.240	0.206	0.143	−0.342
8 BOTTOM (LEFT)	−0.096	−0.500	−0.240	0.206	−0.143	−0.342

If Fig. 13.5 is examined more closely it is evident that each ellipse uses part of its boundary to form part of the circular boundary. This will be referred to as the active boundary of the ellipse. The bold circular 90° arc is transformed to this active ellipse boundary. Each of the above eight transforms is used to form a different one of the 45° arcs that make up the complete circle in Fig. 13.5. However, owing to the fractal nature of the picture, each ellipse is made up of eight more smaller ellipses around its boundary.

For each ellipse, there are two smaller daughter ellipses forming part of the circular boundary. Progressing to higher and higher magnifications, each ellipse is found to be composed of eight smaller ellipses around its boundary. Furthermore, each ellipse lying on the circular boundary has two smaller daughter ellipses which also lie on the circular boundary.

When considering the modification required to plot just the circular boundary, it has been shown that each ellipse is composed of eight smaller daughters. To plot the circular boundary only the two daughters that lie on the circular boundary need to be processed; the other six can be ignored. As described in the introduction, the deterministic algorithm recursively spawns eight new procedures at every stage — one for each affine transform in Table 13.4. The algorithm can be adapted such that only two procedures are spawned at every stage, choosing the two correct daughter affine transforms corresponding to the smaller ellipses lying on the circular boundary. All affine transforms in Table 13.4 have daughter transforms 1 and 2.

The fact that all the daughters are transforms 1 and 2 is not a coincidence. The transforms were originally chosen so this would happen in order to illustrate the use of fractal boundary segments as library calls.

Shown in Table 13.5 are affine transforms 1 and 2 that were used to create the circle. As all the daughters are present this is a proper fractal code and forms the right-hand 90° 'circular' arc. This is a boundary segment. By copying the 90° arc to an adjacent position a 180° arc can be constructed, and then the 180° arc can be copied to complete a circle. Each copy procedure can be performed with one additional transform as shown in Table 13.6. Transform 3 corresponds to a copy and 90° clockwise rotation, and transform 4 to a copy and 180° rotation. Consequently a circle boundary has been expressed in terms of eight ellipses by using only four affine transforms and the additional daughter information.

Looking at another example of a boundary segment, a straight line, the two affine transforms shown in Table 13.7 form a horizontal line passing through $y = \frac{1}{2} = 0.5$. Each half of the line is defined in terms of a reduced and shifted original. By copying the line to the bottom of the image a set of horizontal parallel lines can be constructed, and by copying and rotating through 90°, the two vertical parallel lines can be produced that complete a line drawing of a square. The additional transforms to do this are shown in Table 13.8. Transform 3 copies and shifts the top edge to the bottom, and transform 4 copies and rotates the horizontal edges 90° clockwise to form the vertical edges.

Table 13.5 IFS code for 90° circular arc (east quadrant of circle)

Transform	A	B	C	D	E	F	Daughters
1 NORTH EAST	0.240	−0.206	0.096	0.500	0.342	0.143	1 2
2 SOUTH EAST	0.240	−0.206	−0.096	−0.500	0.342	−0.143	1 2

Table 13.6 Additional transforms to Table 13.5 to give full circular boundary

Transform	A	B	C	D	E	F	Daughters
3 SOUTH	0.0	1.0	−1.0	0.0	0.0	0.0	1 2
4 WEST & NORTH	−1.0	0.0	0.0	−1.0	0.0	0.0	1 2 3

Table 13.7 IFS code for horizontal line (top edge of square)

Transform	A	B	C	D	E	F	Daughters
1 LEFT	0.5	0.0	0.0	0.0	−0.25	0.5	1 2
2 RIGHT	0.5	0.0	0.0	0.0	0.25	0.5	1 2

Table 13.8 Additional transforms to Table 13.7 to give full square boundary

Transform	A	B	C	D	E	F	Daughters
3 BOTTOM	1.0	0.0	0.0	1.0	0.0	−1.0	1 2
4 LEFT & RIGHT	0.0	0.0	1.0	−1.0	0.0	0.0	1 2 3

The above examples are for trivial shapes but the same procedures can be applied to create any complex boundaries. Conversely by breaking down real boundary data into a set of simple boundary segments those boundaries can be encoded using affine transforms.

13.4. ONE-DIMENSIONAL CODING

Recently fractal techniques have been applied to coding real data [1]. This has required a reversal of the approach described above. Instead of using an algorithm with a local rule to create fractal pictures, the data has to be broken up into a set of patterns for which a local rule can be found for generating the image data.

The following block-based fractal coding technique exploits similarities in different parts of a one-dimensional signal. This could be a line of image data for instance, or a speech signal. In the picture of a fern (Fig. 13.2) it can easily be seen where these similarites lie, each fern leaf being a smaller fern. With more general data it is not as obvious where these similarities lie. That is because human perception is dominated by the macroscopic detail in the signal. If the data is examined at a microscopic level by splitting the signal into blocks, each containing four samples, then these blocks show remarkable similarities — blocks that are predominately flat or blocks with a ramp across them, for example.

Consider the example shown in Fig. 13.6. Similar patterns in different parts of the signal must be matched by using affine transforms. As mentioned above, these affine transforms have to be contractive. A method of forcing a spatial contraction is to pattern-match between blocks with four samples and blocks with a number greater than four, for instance eight. The blocks with four samples must tile the signal and are called range blocks. The blocks with eight samples are called domain blocks and can be taken from any part of the signal.

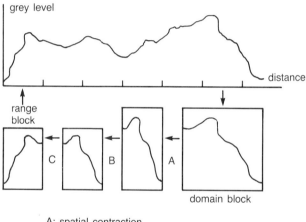

A: spatial contraction
B: grey level contraction and offset
C: geometric mapping (reflection)

Fig. 13.6 Block matching for 1-D fractal coding.

For each range block in turn a domain block must be found that has a similar pattern. Maximizing the number of different patterns that can be extracted from the signal will improve the chance of finding a good pattern match. The following techniques can be used to improve the wealth of patterns.

- A new domain block could start every sample. Although a domain block consists of eight samples, there is no reason why domain blocks cannot overlap.

- The grey-level (amplitude) data in the block can be scaled and/or offset, the scaling parameter being constrained only by the contractivity criterion mentioned above.
- The data within the domain block can be re-ordered, for instance, by reflecting it.

When encoding a signal, the type and magnitude of these data manipulations must be known by both the encoder and decoder. Therefore in order to code a range block, the encoder must transmit codewords to the decoder that indicate the location of the domain block in the signal and also what type of modifiers must be used in order to generate the correct pattern for the range block.

There are many possible domain blocks in a signal, and there are a multitude of modifiers that could be used in order to try to generate a suitable pattern match for each range block. However, all this information must be transmitted to the decoder, so it is prudent to use only the minimum number of domain blocks and modifiers as is necessary to produce a set of patterns rich enough that most range blocks can find a satisfactory match. This reduced set of modifiers is a subset of the affine transform — referred to below as a fractal transform.

In Fig. 13.6 a domain block from one part of the waveform is contracted spatially (A), then its grey levels are re-scaled by a multiplicative factor and an offset (B), and finally it is reflected (C), to provide a match for a range block on another part of the waveform. By performing similar procedures for every range block in the signal, a set of fractal transforms can be found to describe the whole signal in terms of itself.

Having schematically demonstrated the encoding process, the decoding process can be dealt with in a similar way. The decoder is initialized with an arbitrary signal, and parts of the signal corresponding to domain blocks are taken, modified, and used to replace other parts of the signal corresponding to range blocks. The striking aspect of fractal transforms is that, as the above algorithm is iterated, the decoded signal becomes increasingly more like the encoded signal and less like the arbitrary signal used as a starting point.

By explaining Fig. 13.7 it will be seen how this is possible. At the encoder there exists a simple signal consisting of four range blocks (1,2,3,4) and two domain blocks (A,B). Below each range block is a column, listing which domain block and what type of fractal transform has been used to code each range block. This information is transmitted to the decoder. The decoder is initialized with an arbitary signal; the one chosen contains all values set to mid-grey, so both domain blocks A and B contain a flat signal set at the mid-grey value. These are then modified and used to replace the four range

278 IMAGE DATA COMPRESSION USING FRACTAL TECHNIQUES

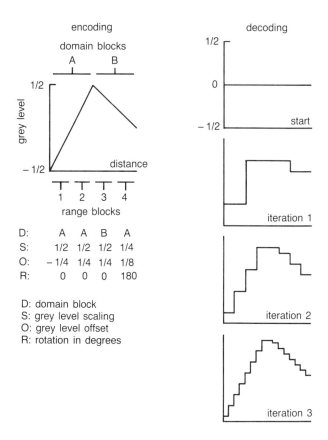

Fig. 13.7 Decoder convergence for 1-D fractal coding.

blocks as dictated by the information given in the columns. For example, range block 1 takes domain block A, spatially contracts by ½, and offsets the grey level by $-¼$. The grey-level contraction of ½ has no effect at this point because the data in domain block A is flat. Inspecting the decoded signal after one iteration it is evident that the signal for each range block is flat, but the signal in the domain blocks now has a step edge halfway across. The new domain block data is now used in the second iteration. The same transforms are performed again; after the second iteration each range block has a step edge halfway across it. However this time the grey-level scaling parameter has had an effect. The step edge in the range blocks is not as high as in the domain block that was used to create it. After the second iteration the domain blocks now have three step edges in them. As the same transforms are iteratively repeated the signal at the decoder converges towards the signal at the encoder. Eventually the step edges will become so small that they will map on to an individual sample in the data, and there is no point in continuing the iterations.

If the starting signal at the decoder had been different, a sine wave for instance, the signal at the decoder would still have converged to the signal at the encoder, although it might take a different number of iterations. This is because only contractive transforms were allowed. That is, in Fig. 13.6, the area of the domain block is greater than the area of the range block. At the decoder any error associated with the original starting signal is reduced by the grey-level scaling factor, at each iteration, until it becomes insignificant.

A practical way of using fractal transforms to code real signal data in one dimension has been explained. In order to code still pictures the above algorithm has to be extended to operate on two-dimensional data.

13.5. STILL-PICTURE CODING

The scheme for two-dimensional coding is very similar to that for one-dimensional coding. The picture is split up into 4×4 range blocks consisting of 16 pixels (picture samples), which perfectly tile the image. Domain blocks are used which are typically 8×8 in size. These overlap such that a new one starts every four pixels in both the horizontal and vertical directions. This technique was originated by Arnaud Jacquin [2].

Consider Fig. 13.8, a schematic of a 2-D grey-scale image. A domain block in such an image can be represented as a 3-D cuboid. Its base is 8×8 pixels, and its height is the grey scale 0...255. A transform converts this cuboid to a smaller range block cuboid 4×4 with a modified grey scale. The modified grey scale is restricted to a range that ensures the range block cuboid is smaller in volume than the domain block cuboid. Owing to this contactive mapping any error associated with starting on a random picture is reduced at every iteration in line with the contraction factor of the transform.

In one dimension it was possible to manipulate the data geometrically, for instance, by reflecting it. In two dimensions eight simple geometric manipulations can be performed, as shown in Fig. 13.9.

280 IMAGE DATA COMPRESSION USING FRACTAL TECHNIQUES

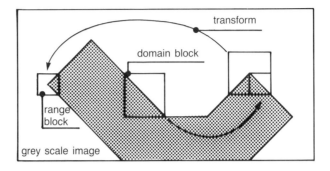

Fig. 13.8 Block matching on picture edges for schematic of 2-D grey-scale image.

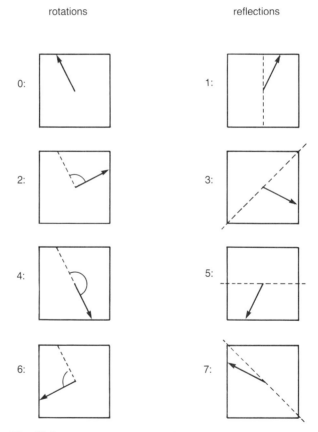

Fig. 13.9 Set of simple geometric manipulations for a square.

The practicalities of implementing this scheme will now be covered. If the decoder is given a blank screen as a starting point, Jacquin's decoder takes about four to eight iterations to converge. If an arbitrary picture is used as a starting point, the decoder will take a few more iterations to converge (see Figs. 13.11 – 13.26). If the starting point is a picture closer to the original then the decoder will converge more quickly. With this in mind, the transform data was inspected more closely.

For every range block there is a grey level offset in the transform parameter. For images with grey levels of 0...255 the offset has a range −255...255. If the range block grey-level means are transmitted for every range block, these offsets do not need to be transmitted, and the decoder has enough information in the range block means to compose a picture remarkably close to the original.

For a picture size 720×576 there will be 180×144 block means. This represents a smaller version of the original picture. At the decoder this small picture is enlarged, using interpolation, back up to size 720×576. This image, which looks like a blurred version of the original, is used as the starting point for the fractal iterations. In practice it is found that after one iteration the decoded picture has converged. That is, the domain blocks are found in the blurred picture, fractally transformed, and placed in the range block positions in the final picture in one iteration.

In this scheme 12×12 domain blocks were used to speed up the convergence. These overlapped, such that a new one started every four pixels both horizontally and vertically, as shown in Fig. 13.10. As can be seen a domain block boundary always lies on a range block boundary. This criterion forces domain blocks to be tiled by range blocks. At the decoder, the grey-level mean of a domain block can therefore be computed from the range block means from which it is tiled. The grey level offset part of the fractal transform depends only on the respective domain block mean and range block mean. As all this information is now available at the decoder, the grey level offset parameter does not have to be transmitted. This can be expressed more formally as follows.

The transformation from pixels grey levels in domain blocks to pixels grey levels in range blocks can be expressed as:

$$\frac{(D_i - U_d)}{S_d} = \frac{(R_i - U_r)}{S_r}$$

282 IMAGE DATA COMPRESSION USING FRACTAL TECHNIQUES

where i is pixel index
 U is mean grey-level of block
 S is standard deviation of grey-level
 d is domain block
 r is range block.

Re-arranging, the range block pixels value is given by:

$$R_i = \frac{S_r}{S_d} \cdot D_i + U_r - \frac{S_r}{S_d} \cdot U_d$$
$$= \text{scaling} \cdot D_i + \text{offset} \qquad 13.1$$

where offset $= U_r - \text{scaling} \cdot U_d$

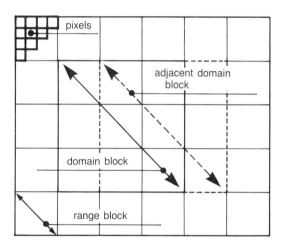

Fig. 13.10 2-D image mapping for 12×12 domain blocks.

As can be seen from equation (13.1), the grey-level offset can be computed at the decoder from the scaling parameter and the associated domain block and range block means, the domain block mean being evaluated from the means of the range blocks used to tile it. However, on further inspection, it can be shown that the offsets do not have to be computed at all.

By transmitting the range block grey-level means, a blurred version of the original picture can be constructed at the decoder. This is essentially a low-pass filtered version of the picture. At the encoder, if the same low-pass picture is constructed and subtracted from the original picture, a high-pass picture is produced. It is the information in the high-pass picture that the decoder now requires.

In areas of low picture activity the high-pass picture will consist of mainly zeros. This information can be economically transmitted to the decoder by classifying range blocks into flat blocks and edge blocks. If the range block information in the high-pass information is sent to the decoder. The other remaining blocks are classed as edge blocks and high-pass information has to be sent for these blocks. The decoder now needs to be told whether a range block is an edge or flat type. This is overhead information but is insignificant compared to the amount of data saved by not transmitting the fractal transform information for the flat blocks.

In the high-pass frame the range blocks all have a mean grey level of zero. As domain blocks are tiled by range blocks, this forces the domain block grey-level means to zero also. The fractal transform grey-level offset in equation (13.1) is therefore zero for all transforms applied to a high-pass frame, and needs no computation.

The domain blocks are taken from the low-pass picture, so at the decoder the mean grey level of the domain block must be subtracted from the domain block data before it is transformed and used to code the edge-type range blocks in the high-pass frame. When the high-pass frame is full decoded it is added back into the low-pass frame to create the final decoded picture.

The above coding scheme was applied to the international test picture 'ZELDA' — used as an ISO JPEG standard picture [3]. This CCIR format picture can be compressed to 0.8 bits per pixel, without introducing artefacts obvious to the untrained eye. Compression down to 0.5 introduces visible artefacts but the picture quality is still reasonable. No quantization has yet been introduced into the scheme, so it would seem possible that a similar coding quality to the ISO JPEG standard could be produced, if more image compression techniques were applied. Results are shown in Figs. 13.27—13.30.

The decoder is written in Pascal and has not been optimized for speed. When allowing one iteration at the decoder, it runs at about 16 μs per pixel on a 5 MIPS VAX 6310 computer. If rewritten in assembler, one would expect at least a two-fold increase in speed. Transferring the code to a 386 IBM-compatible PC with VGA graphics (rated at 2½ VAX MIPS) this would enable a 320×200 picture to be decoded in about 1 s.

Fig. 13.11   'Zelda' original (720×576 pixels, 8 bits per pixel).

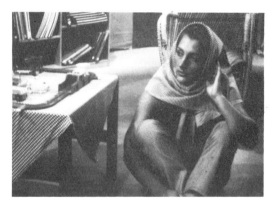

Fig. 13.12   'Barbara with toys' starting picture for Jacquin scheme.

Fig. 13.13   Jacquin scheme after one iteration (approx 2 bits per pixel).

STILL-PICTURE CODING 285

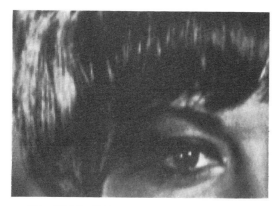

Fig. 13.14 Enlargement of 'Zelda' original.

Fig. 13.15 Enlargement of 'Barbara with toys'.

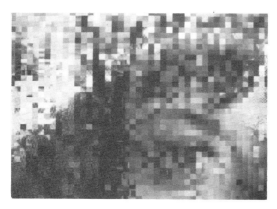

Fig. 13.16 Enlargement of Jacquin scheme after one iteration.

286 IMAGE DATA COMPRESSION USING FRACTAL TECHNIQUES

Fig. 13.17 Jacquin scheme after two iterations (approx 2 bits per pixel).

Fig. 13.18 Jacquin scheme after three iterations (approx 2 bits per pixel).

Fig. 13.19 Jacquin scheme after five iterations (approx 2 bits per pixel).

STILL-PICTURE CODING 287

Fig. 13.20 Enlargement of Jacquin scheme after two iterations.

Fig. 13.21 Enlargement of Jacquin scheme after three iterations.

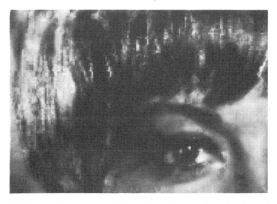

Fig. 13.22 Enlargement of Jacquin scheme after five iterations.

288 IMAGE DATA COMPRESSION USING FRACTAL TECHNIQUES

Fig. 13.23 Jacquin scheme after eight iterations (approx 2 bits per pixel).

Fig. 13.24 Jacquin scheme after twelve iterations (approx 2 bits per pixel).

Fig. 13.25 Starting image for author's scheme (approx 0.3 bits per pixel).

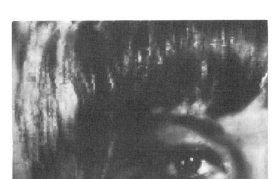

Fig. 13.26 Enlargement of Jacquin scheme after eight iterations.

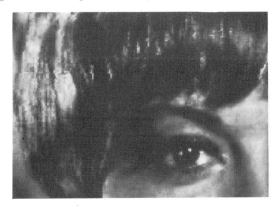

Fig. 13.27 Enlargement of Jacquin scheme after twelve iterations.

Fig. 13.28 Author's scheme after one iteration (approx 0.3 bits per pixel).

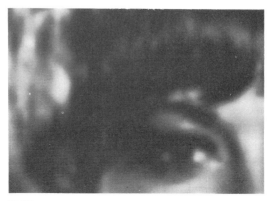

Fig. 13.29   Enlargement of starting image for author's scheme.

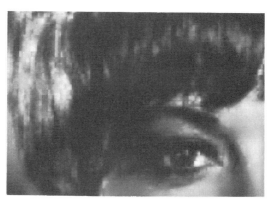

Fig. 13.30   Enlargement of author's scheme after one iteration.

13.6   MOVING-PICTURE CODING

Moving-picture sequences may readily be coded by extending the foregoing scheme to three-dimensional blocks of data; consequently the range blocks now become $4 \times 4 \times 4$ in size, and the domain blocks $12 \times 12 \times 12$. Ideally one would like to take domain blocks from any part of the sequence, but this would mean holding all the frames of the moving sequence in memory. This is not practical, so the sequence is divided into a series of mini-sequences of 'slabs'. All the volume data associated with a slab is held in memory at the same time, and 3-D transforms operate on cubes of data within the volume. Consequently is it possible to use time information to code spatial information and vice versa.

The initial approach, shown as a schematic in Fig. 13.31, was to code each slab as a series of 'thick' still frames. In practice this entailed taking slabs of sequence 12 frames thick, as domain blocks are now of size $12 \times 12 \times 12$. With a frame rate of 10 Hz this corresponds to 1.2 s of moving picture. There are three frames of $4 \times 4 \times 4$ range blocks spanning the time axis of the slab; all three frames have to be coded from the one frame of $12 \times 12 \times 12$ domain blocks before moving on to the next slab. The coding process is done in much the same manner as for still pictures.

Each slab of the original sequence was separated into a low-pass slab and a high-pass slab, the low-pass slab being made up of the range block means. At the decoder domain blocks were taken from the low-pass slab. A prediction scheme was used to reduce the amount of data transmitted to the decoder for the low-pass slab. The range block means were DPCM-coded [4] using predictions from the range blocks in the previous slab, previous row and previous column, the difference between the prediction and the actual value being transmitted. Consequently in the parts of the picture with no moving objects the difference value would be zero. The high-pass slab was coded using fractal transforms for the edge-type range blocks.

Unfortunately, although the data compression for the 3-D coder was impressive, the picture quality was very poor. In the decoded sequences severe blocking artefacts were produced. These were the 3-D range blocks in the high-pass band, and they encompassed a moving edge in time and space. It is worth explaining this effect in more detail, because it reveals some interesting facts about the human visual system.

A decoded 3-D range block spans four frames in time. The first frame of this block is on the screen; the moving edge that it is coding has yet to arrive, but the errors in the block are quite evident against the smooth low-pass background. The second and third frames of the block contain an edge sweeping across the image and the errors in the block are hidden on the spatial edge. The fourth frame appears, the edge has now passed, and again the block errors are clearly visible against the surrounding low-pass background.

The above effect can be reduced by increasing the frame rate at which the image is coded. A 25 Hz version produced the same artefacts but they were constrained closer to the moving edges. As the frame rate is increased each 3-D block of data spans less time, because the number of frames in a block remains constant. As the time spanned decreases, so the time variance of the data decreases. If the time variance was zero the 3-D block of data

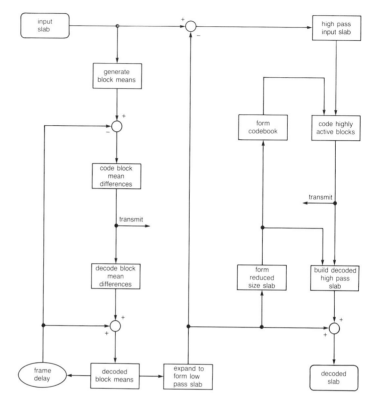

Fig. 13.31 3-D 'slab' image sequence coder.

could be represented as a 2-D block. Hence in increasing the frame rate to reduce the blocking artefacts one is effectively progressing towards a still-frame coder.

A second approach was tried in order to code moving sequences without annoying artefacts. The method finally settled on is essentially the method used for still pictures. Each frame of a 10 Hz moving sequence was coded as if it were a still picture. This type of coding is known as intra-frame coding. However the method is not truly intra-frame, as will be described below.

The first frame in the sequence is a special case and is coded exactly as in the still picture case. For all subsequent frames the method is as shown in Fig. 13.32. Each frame is split into a low-pass and high-pass picture as before, the low-pass picture consisting of the range block grey-level means. These are DPCM-coded [4] using predictions from the previous frame, previous line and previous pixel. The high-pass picture is coded using the same method as for still pictures, except that instead of the decoder taking

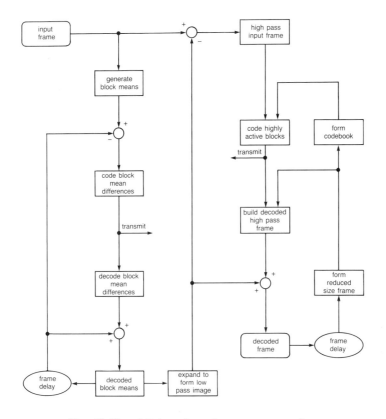

Fig. 13.32 2-D intra-frame image sequence coder.

domain blocks from the low-pass picture they are taken from the previously decoded frame. This gives a slightly crisper appearance to the final decoded picture as there are higher frequencies in a decoded frame than in a low-pass frame.

It became apparent that using the above method and coding at 10 Hz, a 352×288 monochromatic moving-picture sequence could be coded at the data rate of 80 kbit/s with reasonable quality. There were slight twinkling artefacts on moving edges, and a soft focus look about the low-activity parts of the sequence, but the sequence was quite watchable, and without viewer fatigue. Results are shown in Figs. 13.33 – 13.38.

Fig. 13.33 Still from 'Miss America' original sequence (352 × 288 pixels, 30 Hz frame rate, 24 Mbit/s).

Fig. 13.34 Still from intra-frame coded sequence at 10 Hz frame rate (approx 90 kbit/s).

Fig. 13.35 Still from slab coded sequence at 30 Hz frame rate (approx 90 kbit/s).

MOVING-PICTURE CODING 295

Fig. 13.36 Still from 'Casablanca' original sequence (352 × 288 Pels, 30 Hz frame rate, 24 Mbit/s). © 1943 Turner Entertainment Co. All rights reserved.

Fig. 13.37 Still from intra-frame coded sequence at 10 Hz frame rate (approx 100 kbit/s). © 1943 Turner Entertainment Co. All rights reserved.

Fig. 13.38 Still from slab coded sequence at 30 Hz frame rate (approx 100 kbit/s). © 1943 Turner Entertainment Co. All rights reserved.

13.7. PSYCHO-VISUAL CONSIDERATION

From the above results there are a number of observations to be made concerning the relationship between fractal coding and psycho-visual phenomena. It is sometimes convenient for mathematicians to treat time as a dimension much like the spatial dimensions. It is certainly elegant and simple to extend the fractal coding algorithms from one-dimensional coding of speech-type data, through two-dimensional coding of still pictures, to three-dimensional coding of moving sequences. However it is notoriously difficult to explain the properties of the human visual system with a mathematical model.

- Fractal coding works quite well when coding still edges. Sharp edges have high frequencies associated with them. Fractal coding economically codes high-frequency information. When coding edges with fractals, a sharp edge that approximates the original is constructed. The errors due to this approximation are masked by the edge. This is a psycho-visual phenomenon commonly exploited by image coders.

- When coding in 3-D, before and after the moving edge is on the screen the errors due to fractal coding have no edge by which they can be masked. However in 3-D coding another important psycho-visual phenomenon comes into play — the eye's high sensitivity to movement. In order to survive and put into action the 'fight or flight' behaviour most animals have evolved an acute sensitivity to the slightest movement in the environment. In image sequences this corresponds to changes along the temporal axis. Normally these changes are associated with a moving edge. However when using 3-D fractal coding the image changes from low-pass background to a fractally coded block before the edge arrives, and also back again after the edge has passed. This creates an annoying 'blocky' flickering effect.

13.8. THE FUTURE

To date, fractal coding has progressed only as far as other non-intelligent compression schemes, which make no assumptions about the data they are handling. All such schemes are limited in the amount of compression they can achieve by Shannon's information limit [5]. To make an order-of-magnitude leap in terms of image compression, it is widely believed that object-oriented schemes would have to be used.

In an object-oriented coder, data would be sent representing the types of objects in the image, instead of sending data representing grey levels of pixels in the image. To encode a picture of a red mug for example, a parameter for a mug would be required, and a parameter for the colour red; these would be transmitted to the decoder, which would draw a red mug. The problem with this scheme is that the decoder does not know what type of red mug it is, where in the scene it is, and from what angle it is being viewed, etc. However all this extra data can be parameterized if necessary. The point is that only the minimum number of parameters that are important for the application are transmitted from encoder to decoder.

Fractals fit into this vision of the future in the following way. A decoder drawing a red mug would have to use computer graphics techniques; the surface of the mug would be segmented into regions that are homogeneous; the boundaries of these regions would be drawn first. The regions would then be rendered with different textures — in this case, shades of red depending on an arbitary set of lighting conditions. All the schemes required for object-orientated coding are readily achievable with fractal techniques [6,7,8], such as the boundary and region-coding properties of affine transforms.

ACKNOWLEDGEMENT

The photographs appearing in this chapter as Figs. 13.36, 13.37 and 13.38 are taken from the film 'Casablanca' and are reproduced here by agreement with Turner Entertainment Co, Culver City, Ca, USA.

REFERENCES

1. Barnsley M F and Sloan A D: 'A better way to compress images', BYTE Magazine (January 1988).

2. Jacquin A E: 'A novel fractal block-coding technique for digital images', IEEE International Conference on Acoustics, Speech and Signal Processing (1990).

3. Hudson G P: 'The international standardisation of a still picture compression technique', BT Technol J, $\underline{7}$, No 3 (July 1989).

4. Gonzalez R C and Wintz P: 'Digital image processing', Addison-Wesley (1987).

5. Shannon C E and Weaver W: 'The mathematical theory of communication', Univ Illinois Press (1949).

6. Carpenter L C: 'Computer rendering of fractal curves and surfaces', Supplement to Proc SIGGRAPH 80, pp 1—8 (August 1980).

7. Fournier A and Fussell D: 'Stochastic modeling in computer graphics', Supplement to proc SIGGRAPH 80, pp 9—15 (August 1980).
8. Heckbert P S: 'Survey of texture mapping', IEEE computer graphics and applications, pp 56—67 (November 1986).

14

ARTIFICIAL NEURAL NETS FOR IMAGE PROCESSING IN VISUAL TELECOMS: A REVIEW AND CASE STUDY

C Nightingale, D J Myers, J M Vincent and R A Hutchinson

14.1. INTRODUCTION

The advances made in digital technology over the last 20 years or so have enabled many mathematical, statistical and logical tools to be applied to the solution of problems in image processing for telecommunication. Sophisticated image-coding techniques have enabled very high compression to be obtained for both still and moving pictures, and products and services in this area are beginning to emerge. Videotelephony and videoconferencing are now technically and economically feasible and videoconferencing is a reality. Even so some technical problems arising in visual telecommunication remain obstinately difficult or impossible to solve. In particular any function which requires any but the simplest image analysis will be unlikely to provide robust and reliable service in a general application, unless significant new progress can be made.

From a visual telecommunication point of view some important advances could be made, in both quality improvement, cost savings, and enhanced services were it possible to perform certain tasks in feature extraction and image understanding. The model-based coding of Welsh [1], for example, requires that accurate location and tracking of facial features be possible (see Chapter 11). Intelligent surveillance, which could have important implications for visual telecommunications is based on the assumption that scenes of, for example, railway level-crossings can be automatically understood in terms of traffic incidents. Although conventional methods of addressing these problems continue to be developed, any new technology which suggests the possibility of swifter progress must be of great interest. Such a technology, which has been developing over the last five years or so, is 'connectionism' — alternatively called 'neural computing'. Interest in these artificial neural nets (ANNs) has been building for several years in many countries around the world. Recent conferences on ANNs, for example the International Joint Conference on Neural Networks, have had several thousand attendees, and a society — The International Neural Network Society — and several new journals devoted exclusively to the study of ANNs, (e.g. *Neural Networks, IEEE Trans. on Neural Networks* and *Connection Science*) have come into existence in the last four years.

The underlying motivation for the study of ANNs is the consciousness that the brain, both human and animal, seems capable of easily performing many tasks which are either difficult or impossible for conventional serial computers (Von Neumann machines). Examples include the recognition and understanding of speech in humans, the interpretation of natural scenes, and sophisticated methods of control which allow animals ubiquitous movement.

ANNs are sets of simple processors, highly interconnected with weighted connections, and operating in parallel, whose behaviour is inspired by knowledge of the brain's operation. They are usually capable of self-organization either by learning in a supervised mode, in which particular responses are learned by example, or in an unsupervised mode in which regularities inherent in the data may be reflected in the final organization of the connection weights of the network.

It is the object of this chapter to describe neural nets briefly, to note some interesting areas of application in visual telecommunication, and to offer a case study of feature location by neural computation in more detail.

14.2. PRINCIPLES OF NEURAL NETS

14.2.1 The relationship of the brain to conventional and neural computing

The very great differences between the mechanisms of processing in the brain and those of a Von Neumann computer are well known. The Von Neumann computer — shown in simple block form in Fig. 14.1 — possesses a central processor attached to a memory. Instructions and data are fed sequentially

PRINCIPLES OF NEURAL NETS 301

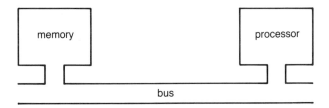

Fig. 14.1 A simplified schematic for a conventional (Von Neumann) computer. Instructions and data are stored in memory, from where they are taken sequentially to be executed in the processor. The results of computations are passed back along.

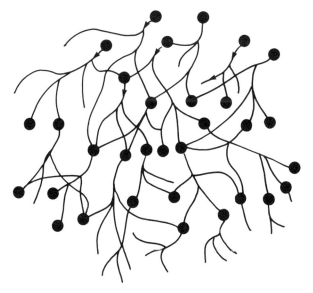

Fig. 14.2 A stylized diagram of a segment of brain. Each neuron receives input from a number of others, and sends its output to yet other neurons. Each neuron can be seen as a processor, with some local memory. It can therefore be regarded as a highly parallel fine-grain computer with the added feature of relatively high connectivity. No contemporary distributed processing system compares with the human brain in number of processors (10^{11}), or number of interconnections (10^{15}).

from the memory into the processor and a program of calculations — including sequence jumps based on logical decisions — is run. The very high speed at which the calculations are now possible enables such machines to perform tasks which are far beyond the capabilities of a human being. For example, a digitized image which may contain several million bits of information can be transformed into the frequency domain in a fraction of a second. The brain, on the other hand, consists of a very large number of neurons, each of which operates relatively slowly — five orders of magnitude

302 ARTIFICIAL NEURAL NETS

more slowly than can a silicon switch (Fig. 14.2). These neurons are connected together via **axons** and **synapses**. The latter, by varying their influence upon the transmission of signals to the neuron, enable learning to take place. Brains may thus learn to perform many tasks without any program of instructions. A child learns to understand speech by constant repetition of words in appropriate situations; he or she has no mechanism by which semantics and syntax could be taught as a sequence of instructions. Yet no computer can be programmed to match the performance even of a five-year-old in this sphere.

As the nature and function of the brain has become better-understood, the power of conventional computers has similarly increased to the point where simulations of highly interconnected brain-like systems have become possible and inevitable. In addition the emergence of new techniques for implementing the high connectivities seen in brains, e.g. wafer-scale integration, optical computing and amorphous silicon have spurred interest in the study of brainlike architectures. Figures 14.3 and 14.4 show a comparison between an artificial neuron and a simplified diagram of a real neuron. The artificial neuron usually takes the form of a processing element

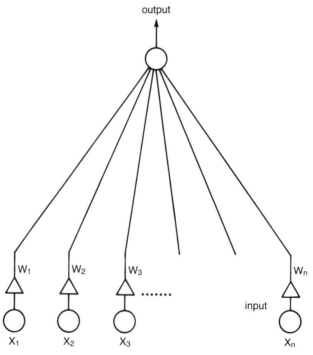

Fig. 14.3 A single-output single-layer neural net, which can be seen as a simple model of a biological neuron.

PRINCIPLES OF NEURAL NETS 303

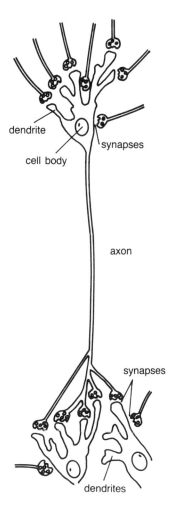

Fig. 14.4 A simplified diagram of the main features of a biological neuron, showing the synapses which modify inputs to a neuron, and enable its behaviour to vary. In the artificial neuron the connection weights, w_i, model these synapses.

with a large number of inputs which are summed to produce an excitation. The single output is then produced by applying a nonlinear **activation** function to this excitation. The activation function often has a monotonic form such that for low values of excitation the output is virtually zero — and for high values it takes a more or less fixed value as shown in Fig. 14.5. It can then be viewed as a device which has an on state, an off state, and a transition zone. Large numbers of these artificial neurons can then be connected

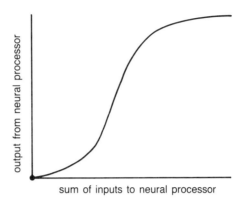

Fig. 14.5 Typical activation function for an artificial neuron: the sum of products of inputs and connection weights is companded with some such function.

together in various clusters and layers to form nets. The connections between them are given weights, which behave like the synapses of the brain and transmit greater or lesser proportions of the signals they carry.

14.2.2 Neural net architectures

The possibilities for assembling neural processors into nets and devising learning stratagems are very numerous, and a large number have been described. Such combinations of architectures and algorithms have been dubbed 'neural net paradigms'. It is intended here, after describing some early paradigms, to concentrate on two, both of which have been studied by the authors and their colleagues, and which in any case act as representatives for the two classes of ANN, the supervised net and the unsupervised net. They are the Kohonen net, which is an example of the unsupervised learning method, and the multilayer perceptron (MLP), which is a good example of a supervised net. The MLP grew out of an earlier idea, the perceptron [2], a single-layer linear threshold net (see Fig. 14.3) which was described and studied for various pattern-recognition tasks and whose capabilities were grossly overestimated.

The single-layer net can be seen as a crude emulation of a neural cell, or neuron, in the brain, as already discussed in conjunction with Fig. 14.3. Its output is formed according to the following equations:

$y = 0$ if $\phi < t$, $y = 1$ if $\phi > T$

PRINCIPLES OF NEURAL NETS 305

where T is a threshold value, and

$$\phi = \sum_{i=1}^{n} w_i x_i$$

The method by which the perceptron is able to learn to classify input patterns into two classes, for example, utilizes an algorithm for altering the weights of the connections, w_1, w_2 ..... w_n during a training phase in such a way that the network eventually gives the correct output in response to the patterns on which it is trained. The training process consists of inputting the desired patterns in sequence, and using the delta (or Widrow-Hoff) rule:

$$\Delta w_j = \eta(t - \sum_i w_{ij} x_i) x_j$$

(where η is the learning rate and t is the desired output) to alter the connection weights.

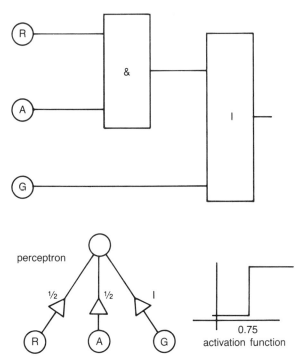

Fig. 14.6 A comparison of a logic circuit and a simple single-layer threshold net is shown in the figure. It is desired to design a circuit which would give a unity output when either a green light, or a simultaneous red and amber light are shown.

It may easily be verified that such a single-layer net can learn to perform certain tasks. For example, if the red-amber-green traffic light signals are used as input, with 110 and 001 to correspond to 1 for advance, and 010 and 001 to correspond to 0 for stop, only three complete cycles through the training data are required to produce a correct response from a set of weights initially set to zero. A comparison between a simple perceptron and a digital logic circuit is shown in Fig. 14.6.

A thorough mathematical analysis of the capabilities of perceptrons was made by Minsky and Papert [3] in their book Perceptrons. They elegantly demonstrated that several important tasks in pattern recognition, for example determining whether a pattern was connected in the topological sense, were beyond the abilities of a perceptron. Worse than this, from the point of view of the ANN researchers of the day, they showed that even a pattern as simple as the exclusive-or logical function could not be learned by a perceptron. The effect of this negative study was to kill most of the interest in ANNs, and to divert funding away from neural nets and into artificial intelligence based on conventional predicate logic.

14.2.3 Multilayer perceptrons

Minsky and Papert's objections were applicable to any linear net, even those with several layers. In addition, it was concluded that single-layer nets could not have their general capabilities increased by introducing a nonlinear function at the output, however complicated its characteristic might be. They also pointed out that although nets that were not subject to these restrictions might not suffer the same functional limitations, there was no reason to suppose that a learning rule, analagous to the delta rule, could be found that would be guaranteed to converge as was the case with the delta rule for perceptrons. In spite of these pessimistic prognostications, multilayer feedforward nets with nonlinear activation functions at each node were studied by a few researchers, since they could be seen to possess sufficient complexity to be outside the scope of Minsky and Papert's theorems on the functional limitations, and a learning rule was discovered which formed the basis of the multilayer perceptron. Although a convergence theorem does not exist for this **generalized delta rule**, the MLPs which are based upon it have been shown to be capable of learning very general tasks, and in practice have solved some practical problems [4]. An example of an MLP is shown in Fig. 14.7. Of course the net could have more nodes and more layers, but the principle of feedforward only is always maintained. If we imagine a typical node in such a structure to be represented by node j shown in the figure, where the weight values associated with the connections confluent upon node

j are represented by w_{ij} ($i = 1,2,...n$), the O_j, the output from node j will be given by:

$$O_j = f(w_o + \sum_{i=1}^{n} w_{ij} x_i)$$

where

$$f(x) = \frac{1}{1+e^{-x}}$$

where w_{oj} is a constant bias at node j.

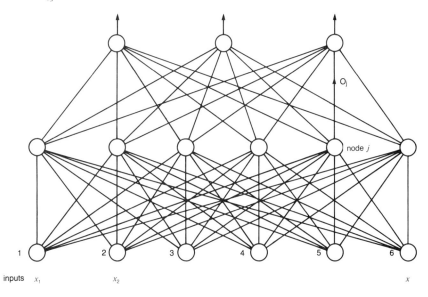

Fig. 14.7 An example of a single hidden-layer multilayer perceptron. Weight values from input node i to node j are given by w_{ij}.

Thus each typical neural processor performs a nonlinear function of its net input. The principle of the learning rule is by back-propagation of the error through the net. The errors E, which are formed of the differences between the teaching signal and the net response at the output layer, are fed back to the previous layer and a set of new error-like parameters formed which are in turn fed back to the previous layer. At each stage a set of corrections for the weights attached to the connections which feed the inputs to a layer are calculated. MLPs have been studied and applied to some of the image-processing and vision problems encountered by the authors and their colleagues, and progress is reported in a later section.

14.2.4 Kohonen nets

The feedforward nets described above, both single and multilayer, have in common the property that a teaching function is applied at the output, and this enables the trainer of the net to arrange for it to give the responses which are required when it has been trained. A number of other neural nets have been described which do not possess this property. A well-known example of the unsupervised net is the Kohonen net [5].

A Kohonen net with an image input is shown in Fig. 14.8, and the principle of operation seen in terms of the application of the net to an image. Suppose a Kohonen net were to be applied to 4×4 blocks in an image, for example. The pixel layer P is connected to a net K in such a way that each pixel node is connected to each element in the two-dimensional layer K. Each of the connections has a weight value associated with it, and some measure of difference between a pattern of pixels and the connection weights by which the pixels of the pattern are associated with a given node exists. A simple measure could be the normal measure of distance in an N-dimensional space, where one point is represented by the co-ordinate vector of the pattern of pixel values, and the other the co-ordinate vector of the set of weights connected to a particular node in the K net layer. Each element in the K layer can then have a level of excitation attached to it for each pattern which is a measure of the closeness of its weight values to that pattern. The Kohonen net then has a means by which a competition amongst all the elements of the K layer results in one being selected as the winner for a particular input pattern. The weights associated with that element, and those in a neighbourhood surrounding it, are then altered in such a way that they more closely resemble the values of the input pattern, thus causing an even greater likelihood that that pattern, and similar patterns, excite that node, or its neighbours. Training then consists of presenting the net with a large number of patterns. At no point in the proceedings is any preconceived idea of the desired output injected into the calculations. The proponents of this type of net regard it as having biological plausibility, in that the brain is indeed structured in a number of layers, with a large number of connections between adjacent neurons in each layer, as in a Kohonen net, and a smaller number of inter-layer connections. Neurons in the retina, for example, excite one another in small neighbourhoods.

The selection of a two-dimensional layout for the K net is an arbitrary one. A three-, or higher-dimensional array could be allowed, or a single line of elements. If the possible patterns on the input are considered as vectors in a 16-dimensional space, for example, then each pattern can be seen as a point in this space, and the totality of patterns in the training set may

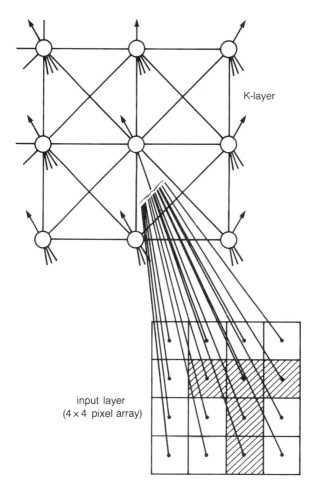

Fig. 14.8 Input to a Kohonen net from a 4×4 array of pixels. The connections to only one node in the K-layer are shown, but each node in the K-layer is similarly fully connected to all nodes in the image array. Connections amongst neighbours in the K-layer are also shown.

form structures themselves, a simple case being, for example, a one-dimensional line of points twisting through the high-dimensional space. If the patterns form some two-dimensional surface-like structure, embedded in the 16-dimensional space, then the distribution of the pattern vectors on the surface will be represented in the K net by a topologically well-behaved model of that surface, and similar patterns will cause neighbouring elements of the K net to become excited. Although it might be reasonable to wonder why it may be supposed that groups of patterns of interest to a human observer have this inherently two-dimensional structure, many examples of Kohonen nets which achieve useful results are known, both in image processing and speech processing [6].

14.2.5 The relationship between neural architecture and the learning function

Neural nets are often thought of as architectures for constructing new forms of computing machinery, but this view sometimes obscures the true nature of the research in the field. The functional feature of the brain, which seems to enable it to solve the problems at which it excels, comes from its learning ability. Whether it is an immediate problem of control, as in the human ability to walk upright, or in a higher function such as interpreting the characters on a page, a task is learned over a period of time. If this learning ability could be emulated, whether by an ordinary Von Neumann machine or by some elaborate new technology, the result would be the achievement of the aim. Of course in searching for the method of reproducing the brain's learning function the brain has been seen as a useful model, and it is not surprising that many successful systems behave like interconnections of neuron-like processors. Such a system can be operated on a conventional computer, and the same functionality attained as would be the case if some special hardware were built. The algorithms are, by their highly parallel nature, very unsuited to the Von Neumann architecture of conventional computers. In order to implement them in such a way as to emulate the speed of response of the brain, special hardware needs to be constructed, to implement true neural nets.

14.3. NEURAL NETS APPLIED TO IMAGE PROCESSING

Image processing and pattern recognition have been an important area of application of neural nets since early days. Neurally inspired machines, like WISARD [7], have already been used in commercial applications, and complex multilayered neural architectures like the Neocognitron [8] have made advances in important areas of potential application. It is not possible to do full justice to the many contributions to this fertile area of research in a short chapter such as this. One difficulty arises with the very varied nature of both the types of net and the applications, in that it is very rare to be able to compare like with like when examining the work of different authors. A feature of the research programme described in this book is the concentration of developments on a restricted problem area. The recognition and location of features in a face has strong motivation in visual telecommunications, for improving the quality of videotelephony images, and it is appropriate to have concentrated on this problem, both from a comparative point of view, and from the point of view of visual telecommunications. There are of course other visual telecommunications applications for neural nets than feature extraction. Noise reduction and image compression are also areas where work has been done. These topics are briefly discussed, and the reasons why there is no neural effort in image synthesis comparable to that in speech synthesis are given. The concentration on visual telecommunications aspects has of course led to many other fascinating and advanced research topics being omitted.

14.3.1 Image coding

Image compression and coding has been an important theoretical and practical field since the possibilty of digital transmission of images became feasible. Coding algorithms and techniques are well advanced, and it is likely that they already achieve near-optimal performance, if techniques in which no higher level knowledge of the picture content is employed. For this reason, although interesting work has been done in conventional image compression using neural nets, it has not been a central theme of neural computation. The work on conventional coding that has been done has used various neural approaches. The hidden layers of an MLP can be used to store an internal representation of an image block, as in the work of Cotrell [9]. Another approach is to use the associative memory aspects of neural nets to achieve compression, by associating a smaller transmitted vector with each input vector. This effectively emulates a well-known image-compression technique, vector quantization, in which a codebook of vectors is searched to yield a representative vector for each block in an image. In a recent study [10] a number of neural techniques and a conventional technique were tested on an image-compression problem. The conventional technique used was the LBJ vector quantization algorithm [11], which was compared to an MLP, a modified Kohonen net and some related nets. In the case of the MLP, training was performed by autoassociating 4×4 blocks of an image, in training, and using the hidden layer, which consisted of either four or two elements, to encode the information. The hidden unit data can then be transmitted, and the layer of weights succeeding the hidden layer can decode the incoming hidden layer information to produce an image block. The process of training the MLP is more computationally intensive than the LBG process for designing a vector quantizer, but the encoding process for an MLP is much less intensive. As far as decoding goes the LBG decoding process is merely a look-up table, and it is therefore simpler than that of the MLP. In the case of the Kohonen net the normal unsupervised process can be used to train a neural net to cluster sets of blocks in an image. The best representative for the cluster will be reflected in the weights connected to the output unit. When the net is in use as a coder, the unit which fires in the Kohonen layer indicates the codebook code to be transmitted, and the representative block may be displayed. The computational requirement in all phases of the Kohonen net is similar to that of the LBG algorithm. The Kohonen net performed better than the conventional LBG coder, both as a stand-alone coder, and as a building block for an adaptive transform coding system — which was superior to all the other techniques. The backpropagation algorithms gave inferior performance to both the Kohonen net and the LBG algorithm, but there was evidence to suggest that the performance could have been improved by more optimization of parameters. Considering the small amount of effort that has been expended in this area results have been very promising.

14.3.2   Image enhancement

An important class of image processes is that of image enhancement. Images may be impaired by the addition of various forms of noise. The noise can arise from the original source, or from processes of transmission and encoding and decoding. One problem with attempting to transmit only changed information in an image sequence is that of distinguishing between pixels which have changed from one frame to the next because of these typical random fluctuations associated with electronic imaging and those due to real changes in the transmitted scene. If noise is transmitted, not only does it constitute an impairment in itself, but it uses transmission capacity, and therefore either increases the cost or reduces the quality of the transmitted pictures. This problem also arises with still pictures.

Neural nets have been suggested as a possible means of improving noise reduction in images. In particular MLPs can be used to filter images; training sets can consist of impaired images as the input set, with noise-free images as output. When the backpropagated error is based upon mean-square error the nets usually approximate classical filters when trained, although no signal theoretical considerations are needed in their development. In one approach the input data consists of a moving sequence with added noise and the output data consists of the original sequence without the addition of the noise. The net then requires two input frames, each corresponding to the same part of the image but in successive frames. Training can consist of scanning the two input frames over successive frames of a moving sequence, and the required output array over the second of the two corresponding frames of the uncorrupted sequence. After a sufficient training period the net may then be tested on a different part of the same sequence and its performance determined using mean-square error criteria in addition to subjective criteria. Wright [12], for example, has performed some experiments in which neural net operators trained on natural images and sequences of images learned spatial or spatiotemporal filter functions.

14.3.3 Image synthesis

Image synthesis may be likened to text-to-speech synthesis in some ways, although very significant differences exist which mean that in effect there is little reason to suppose that neural computation has an immediate role to play in synthesis. The strict analogy with image synthesis would be with sound synthesis, since speech is only one possible type of sound. The production of speech, which is a very complex and subtle form of sound, by rule-based algorithms, has so far resulted in synthetic speech which is machine-like and often barely intelligible. Because of this, neural algorithms have been tried in the hope that by training on human speech, human-like speech can be produced, which has so far not been the case. The production of certain other types of sound — some musical instrument sounds, or sounds of natural phenomena — has been more successful, and such has been the case with image synthesis. We are all familiar with some of the very realistic images that can be generated by computers using geometrical models with ray-tracing and other methods of rendering. Even the slightly unreal textural qualities of the various reflectance models can be a pleasant rather than disturbing effect. And with some newer techniques, such as fractal modelling of mountains, clouds and certain types of plant growth, even this problem is diminishing. Few can doubt that within a few years it will become either very difficult or impossible to recognize that many realistic scenes are in fact synthetic. From this we see that rule-based techniques are perfectly adequate for synthesizing most images, so there seems little point in introducing neural learning algorithms to the field.

14.3.4 Image analysis

The case is very different in image analysis. There are a great number of capabilities in automatic image analysis which have long been sought. There are obvious military requirements for methods whereby targets can be automatically recognized and tracked. The presence of significant installations on large satellite images could ideally be automatically recognized. Civil applications also abound. Methods whereby medical images can be analysed for tumours or other pathological manifestations can aid in diagnosis. Industrial robots can benefit from the ability to identify and locate important aspects of their environment. Quality inspection of certain types of product could be automated using visual information, either by itself or in combination with other techniques. There are also applications in visual telecommunications. Surveillance could benefit from intelligent interpretation of scenes. Model-based videophone coding requires that heads can be identified and located, and that the features of a face can be identified and their positions tracked. Automatic face recognition requires similar capability. And in the long term, if images of all kinds could be automatically characterized, new opportunities in classifying and relating the increasing number of electronically stored images would arise.

From the above we see that image analysis is a large and potentially fruitful area of study. Many great minds have addressed themselves to the problems of vision in the past, including many famous scientists and philosophers — Isaac Newton and Ludwig Wittgenstein being two examples. In their day though, the capability of testing hypotheses did not yet exist. Today there is a very large international research effort in this and related areas — see for example [13]. Pattern recognition, computer vision and image feature extraction have been the objects of intense and thorough study. Yet in spite of the depth of knowledge, the complexity and ingenuity of image analysis systems, and the very powerful processing capability currently available, it is doubtful whether anyone either inside or outside the field would claim that generally speaking the problems of image analysis are near to being solved. There are even suggestions that work in the area of machine vision has to some extent failed — see Jain [14], for example, who also notes that the lack of standard test problems has led to difficulties in assessing the true worth of some techniques. *Ad hoc* solutions of certain restricted problems do exist, but the general problem of finding even fairly simple shapes and features in many images remains unsolved. Although there are many things that can be done with a computer that are quite beyond the powers of the human brain to solve directly, as far as vision is concerned the human brain and many animal brains are still immeasurably superior to computers.

Some reasons for the failure of many computer algorithms are simple. Although we recognize certain features in images which are self-evidently important for understanding them — boundaries of objects for example — their existence is often not deducible from simple measurements, such as edge detectors. Similarly many features that are not important for understanding images — irrelevant shadows for example — constantly give false responses to any measuring scheme designed to detect useful and relevant features. Variations in lighting and tone are constantly mistaken for evidence of nonexistent objects, or lead to the suppression of features which would help to identify objects. It is not the purpose of this section to examine exhaustively the reasons for the failures of many image analysis programs, but a good example can save a lot of explanation. If someone were asked how to recognise an eye, he or she would be likely to give a good pictorial description of an eye: a petal-shaped region surrounding a lighter background in which a blue, grey, green, brown or even violet disc representing the iris seems to float. In the centre of the disc is another concentric disc, black this time, representing the pupil. And above the eye a dark horizontal streak representing an eyebrow. Yet if the same person is then shown a snapshot of a family group on the beach, in which each head may be only a half square centimeter in area he or she will unerringly point to the eyes in the image, even though in all probability none of the features previously suggested as characteristics will be visible in the tiny speck-like eyes. And if the image were to be searched for shapes which resemble these specks by any reasonable measure of similarity, there is little doubt that dozens would appear in the shingle or other inappropriate part of the image. It immediately appears that the eyes were only spotted by their positions in the faces. Yet any algorithm designed to find the faces in such an image would almost certainly involve some way of using the eyes as feature characteristics, and we are led to the conclusion that the operation of eye finding involves unravelling a high degree of complex and variable interrelationships between high- and low-level features.

Because of the poor performance of many rule-based algorithms, and because of the apparent excellence of biological vision systems, it has been very natural to look to the brain as an inspiration for more successful methods for analysing images. Biological vision systems are being studied both from the neurophysical and psychological viewpoints and computational models of the human vision system have been proposed: see [15] for an example. Such models are often complex with several levels of processing and with different representations at different scales. Some of these models have been essentially neural in nature, and there is a very strong relationship between computational models of vision and neural nets. Other computational approaches to practical image interpretation may be strongly rule-based, but it is often true, as in the case of eye location mentioned above, that the rules become extremely complex and require the simultaneous use of different scales if good recognition is to be obtained. The difficulty of discovering the rules that govern such an intuitive and unconscious human capability inevitably draws research in the direction of learning implicit rules by example. Taken as a whole these considerations mean that vision and image analysis have been an early and very significant application for neural nets. Indeed neural approaches in this field are at least as numerous, and possibly more numerous than in any other field, and it is beyond the scope of this book to review them all. It may be of greater value to compare different neural approaches to a limited problem, to try to obtain some measure of success — the more so since lack of standard experimentation has been cited as one of the causes of the limited progress so far attained in the field of machine vision [14].

14.4. PRACTICAL APPLICATIONS

Neural nets have been applied to a number of problems in visual telecommunications, and two such applications are considered here. Intelligent surveillance, in which the necessity for constant human monitoring is obviated by a computer vision system capable of identifying significant events in a scene, may be undertaken using neural nets. In addition some methods of achieving very low bit-rate transmission of videotelephony require significant image analysis which has been carried out using a hierarchy of agents of which some or all may be neural.

14.4.1 Surveillance

Neural nets have recently been receiving much attention in the field of video-based surveillance. Scenarios in which video-based systems are attractive generally fall into one of two categories: those involving long periods of inactivity which may be sporadically punctuated by significant events and those requiring repetitive classification of a limited set of 'objects'.

The first group is typified by crime investigations: these often involve the surveillance of a particular scene for days on end, whilst awaiting the occurrence of a particular event such as entry to a building. More commonly this group is represented by non-investigative security surveillance of, for example, industrial sites or prison compounds. In such situations a security guard would typically be responsible for a number of cameras, each surveying a different area or view of the site, and being viewed on a single monitor in sequence. The guard's attention may be focused on the input from a particular camera through an alarm triggered by a simple motion-detection system; such systems are regularly falsely triggered by unimportant, or legitimate, motion within the scene, e.g. a tree blowing in the wind, birds and animals, or even the shadow of a cloud scudding across the sky. A system capable of detecting, extracting and classifying objects would clearly be beneficial in reducing the number of false triggers, hence improving on the security and making more effective use of the security guards time.

The second set of scenarios referred to above is typified by traffic monitoring. Statisticians often require information relating to road or junction usage, broken down into vehicle types; currently this requires a number of people to count vehicles laboriously over a long period of time. Once again it is apparent that a video-based system capable of classifying and counting different types of vehicle would be of great assistance in such situations [16], [17].

A number of workers have demonstrated rule-based systems for traffic monitoring [18]. It is a relatively trivial step to extend such systems by, for example, incorporating a neural net, to perform the necessary classification. One such system, described here, has recently been developed and applied to unmanned railway level crossings; the techniques are, however, equally applicable to many of the scenarios outlined above. The object detection and extraction are conventional pixel-level operations: incoming frames are subtracted from a reference background frame to produce a difference frame, thereby highlighting any changes in the scene content. The difference frame is thresholded and filtered to remove noise and highlight the location and shape of any important objects; these may then be extracted and applied to a neural net for classification.

The image data which has been extracted must be pre-processed such that it is in a suitable format for input to the net. One possibility is simply to scale the binary representations of the detected objects to a standard input size and apply the pixel information directly to the input of the net. In preliminary tests with limited data sets the author has shown that typically, when classifying objects on a level crossing into one of four categories — large vehicles (e.g. lorries), small vehicles (e.g. cars), bicycles/motorbikes and people — 50% correct classification can be achieved, based solely on such binary shape information. When the scale information, or to be more specific the area and aspect ratio, is now included the percentage correct classification has been shown to approach 100%. Precisely how the scale information is incorporated, in particular with a realistic weighting, is itself a question of some complexity: empirical evidence suggests that one may simply multiply particular inputs to a net by some weighting factor to increase their relative importance. A theoretical basis for this assumption has yet to be developed. An illustration of boxing a vehicle, together with the resulting histograms, is shown in Fig. 14.9.

As an alternative to using binary image data one may process the image so as to extract lower-level parameters and use these as the basis for classification. One such approach is to develop a reduced set of Fourier descriptors which describe the boundary of the detected object [19]. Provided that one can extract valid object boundaries (which is in itself a non-trivial task) this approach offers a number of advantages: for example, the number of inputs required by the neural net, and thus its complexity, is vastly reduced; and the scale information may be made implicit in the Fourier descriptors,

318 ARTIFICIAL NEURAL NETS

Fig. 14.9 Classifying objects in a level crossing scene using neural networks.

so that one does not have subsequently to recombine this information. Preliminary results indicate a high correct classification success rate.

14.4.2 Image compression for low-bitrate videotelephony

Currently, the most important aspect of image processing in telecommunications is the possibility of improving methods for compression of image data. Uncompressed digitized television pictures may require a transmission channel with a capacity of as much as 216 Mbit/s, which would require thousands of ordinary telephone channels to sustain; this is prohibitively expensive. Consequently there has been a continuing drive towards techniques which reduce the amount of data which must be transmitted, whilst retaining as high a quality as possible. Noise reduction, as described in a previous section, plays an important role in the process of image compression. In addition powerful redundancy removal methods are needed, which must be applied in the space domain for still images, and in both the space and time domain in moving images. Good results have been obtained using algorithms based on the particular statistical characteristic of images as data. International standards for the transmission of still pictures, as well as moving images, at particular bit-rates have been established

— at 64 kbit/s for example. In the case of moving pictures the recognition of moving areas of the picture, and the prediction and interpolation of frames from existing frames, have been an essential part of achieving so much compression whilst retaining adequate image quality. At present for rates as low as 64 kbit/s it is usual to transmit a reduced number of frames so that pictures can appear jerky. It is also usual to reduce the requirement for data to be transmitted by using the current frame with some motion estimation to predict the next frame, and transmit only the error between the prediction and the true frame. A method using a synthetic model head whose rotations, translations and mouth movements can be transmitted as simple instructions can form the basis of a very low bit-rate videotelephone — see Chapter 11. It can also enable better approximations to a true frame to be used to insert missing frames, reducing jerkiness, and to make more accurate predictions so reducing the transmission rate — or improving quality.

If such a model is to be used, it is essential to both locate and track features in a human face, so that correct movements of the face and features can be transmitted to the receiver.

14.5 NEURAL NETS FOR TELECOMMUNICATIONS: A CASE STUDY

14.5.1 Background

The remainder of this chapter describes work recently undertaken to produce a system, based on the use of neural network feature detectors, to locate and track features robustly in digital image sequences. The work concentrates on the location of features in a human face, in accordance with the requirements of section 14.4.2 described above. However, the techniques described should be applicable to determining the position of localized features in general images.

In section 14.5.2 initial experiments are described, which were performed to assess the feasibility of using neural networks as feature detectors in grey-scale images, and the results of these experiments, which lead to the formulation of the hierarchical perceptron feature locator (HPFL) architecture, are discussed. Section 14.5.3 describes the components of the HPFL system at its present stage of evolution and section 14.5.4 describes an envisaged hardware implementation. Conclusions and avenues for further development are discussed in section 14.5.5.

14.5.2 First experiments

Initially, a number of experiments were performed to assess the feasibility of using neural networks as feature detectors in grey-scale images. The MLP, trained using the backpropagation algorithm, was chosen as the neural net paradigm to be investigated, although some experiments were conducted using Kohonen self-organizing nets [20]. Other workers have applied alternative neural paradigms to the same eye-location problem. Stochastic nets have been used by Bishop [21], radial basis function nets by Debenham [22] and self-organizing nets by Allinson [23].

Most experiments were performed on data taken from a set consisting of 60 human head-and-shoulder grey-scale images, 30 male subjects and 30 female, facing the camera and at approximately the same scale. The images were available at a number of resolutions, including 256×256, 128×128, 32×32 and 16×16 pixel images, each pixel consisting of eight bits. The lower-resolution images were formed by pixel averaging and subsampling the 256×256 pixel image. All the images used had associated with them data defining the location of the eye, as determined by a human operator.

The experiments concentrated on locating one eye in the case of higher-resolution images, or both eyes in the case of lower-resolution images. The form of all the experiments was similar: a $j \times k$ pixel window, which constitutes the input to the MLP, was raster-scanned a pixel at a time across the $n \times n$ image. The arrangement is shown schematically in Fig. 14.10. The MLP had $j \times k$ inputs and a single output node, and was trained to output a high value (1.0) when the input window is centred on the eye, the response tailing off over a few pixels to produce a low value (0.0) when the window is displaced from the centre of the eye. The values of j,k and number of hidden nodes were varied in different experiments. In all experiments MLPs with at most one hidden layer were used.

In the case of higher-resolution images a $p \times q$ search area ($j<p<n$, $k<q<n$) was used to restrict the search window to the vicinity of the feature of interest. In this case, the number of candidate eye points was reduced to $(p-j) \times (q-k)$. This was done in order to reduce training times, and to increase the frequency of presentation of eye data to the MLP, early experiments having quickly shown that training the MLP by raster scanning across the whole image produces very poor results. This is due to the infrequency of presentation of eye data to the net. In other experiments on lower-resolution images the frequency of presentation of eye data was increased by alternately presenting raster scan data and eye data to the net. Alternatively a small number of 'representative' input vectors were chosen, including the windows centred on the eye. These three approaches to emphasizing the significance of the eye data represent three examples of a

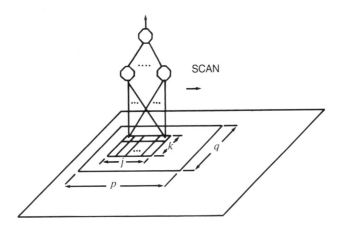

Fig. 14.10   Method used to input image data to the MLP.

wide range of 'probability of presentation' distribution functions that could be envisages.

A number of experiments were carried out, the more significant of which are described below.

14.5.2.1   Lower-resolution feature detection

Preliminary work focused on eye location in lower-resolution images [24]. For 32×32 pixel images an input window size of 7×5 ($j=7$, $k=5$) was estimated to be an appropriate size to cover the eye adequately. An MLP with no hidden nodes was used (i.e. a single-layer perceptron). In the first instance ten training vectors were chosen from a single image, including the two vectors centred on the eyes. The net was successfully trained using this training data set, with expected values of 1.0 for eye vectors and 0.0 for non-eye vectors. Convergence was defined as being when every pattern in the training set produced a network output within 0.15 of the expected value.

When tested by scanning across the image from which the training set had been selected, the net successfully located the eyes. However, although positions of the input window corresponding to eyes resulted in the largest net output values, spurious large values were also generated in other parts of the image. When scanning across an image that it was not trained on, the net generally gave a significant response at the locations of the eyes, although these were not always the highest responses that the net produced.

In an experiment on an image sequence ('Miss America') the sequence was subsampled to form a sequence of 116 16×16 images. The eye positions were manually located and the first five frames were then used to train an MLP with a 5×5 input and no hidden nodes. Training was performed with expected values as for the experiment on 32×32 pixel images described above, but in this case the images were raster scanned in turn to provide input data. The net trained successfully (which is perhaps surprising in view of the fact that only 1 input vector in 60 represented an eye). The trained net was then used to detect eye locations in all the frames. The net proved to be effective in locating eye positions in the sequence, although they were not always located and spurious high outputs did occur, most notably in the mouth and hairline areas. Figure 14.11 shows a frame from the original high-resolution sequence with the MLP output superimposed (white squares).

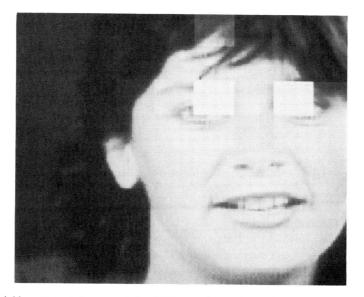

Fig. 14.11 Output of low-resolution MLP superimposed on a frame of the original 'Miss America' sequence (white squares indicate high outputs of the MLP).

14.5.2.2 Higher-resolution feature detection

For 128×128 pixel images, experiments have been performed using MLPs and, for comparison, template-matching techniques [20]. Kohonen self-organizing maps were also investigated but these experiments are not described here; see [20] for details. In all experiments 16 images (eight male, eight female) selected from the data set of 60 images were used to form a training set, and the remaining 44 images were used as a test set. An eye was considered to have been successfully located if the position determined by the technique was within two pixels of the manually determined position.

Template matching

For template matching an input window of 16×16 was used. The template was considered to be over the eye when the pixel to the upper left of the middle of the template was centred on the eye. Two approaches were used to form the templates. In the first approach 16 different templates were used, based on 16×16 ($j, k = 16$) windows centred on the left eye of the training set (in all discussions of image data, left and right refer to positions in the image as seen by the observer). Each of the templates was scanned over a 32×32 search window centred on the eye ($p, q = 32$) in each of the images from the test set. At each position the sum of squared differences between the pixel values in the template and the image was evaluated. For each image the template position with the lowest sum of squared differences, for all the templates, is considered to be the location of the eye.

In the second approach, a composite template was created by taking the average of the pixel values at each pixel position. This was used in a similar manner to the previous template method.

The 'best of 16' approach successfully located eyes in 29 of the 44 images in the test set, a success rate of about 66%. Using the composite template resulted in the location of eyes in 36 of the test set images, a score of about 82%. However the composite template was only successful in locating 7 of the 16 eyes in the training set, leaving its apparently good performance on the test set open to question.

Multilayer perceptron

In the MLP experiments an input window size of 16×16 ($j, k = 16$) was used. Hidden layers of two different sizes were tried; 16 neurons and 32 neurons. The MLP was trained by scanning it across a 32×32 ($p, q = 32$) window centred on the left eye of each of the images in the training set. When the input to the net was centred on the eye the net was trained to output a high value (1.0). The expected value decreased linearly to zero (0.0) at a distance of 6 pixels from the eye centre. Thus near-eye vectors were expected to give a non-zero response. The net was deemed to have converged when all actual values were within 0.1 of the expected values.

The network was first trained on four images of the training set, until it performed eye recognition satisfactorily on these images (although it did not attain the convergence criterion). It was then trained further on eight images including the initial four, and finally on all 16 images. This was repeated for MLPs with 16 and 32 hidden nodes. The best results obtained for the MLP with 16 hidden nodes were a recognition of 38 of the 44 test images, a success rate of about 86%. The MLP with 32 hidden nodes located the left eye in 39 of the 44 test images, a success rate of about 89%. Results for all the approaches to eye location in 128×128 pixel images are summarized in Fig. 14.12.

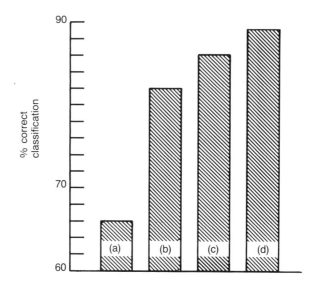

Fig. 14.12 Scores for eye location within 2 pixels on a 128 × 128 image: (a) best of multiple templates, (b) average template, (c) MLP with 16 hidden nodes and (d) MLP with 32 hidden nodes.

Experiments have been performed on 256 × 256 pixel images [25]. In these experiments an input window of 30 × 16 ($j = 30, k = 16$) was used, with a hidden layer of 32 neurons, and a search window of 46 × 32 ($p = 46, q = 32$). The expected output values were similar to those for the 128 × 128 pixel experiments, but scaled to the large image size. Also, an eye was considered to have been successfully located if the maximum response of the net was within 4 pixels of the manually determined eye centre. It must be pointed out that this criterion for success is much less demanding than that in the case of the 128 × 128 pixel experiments; in that case the selection of a random point in the search space would have yielded a correct classification score of 6.25%, as compared with 25% for the 256 × 256 pixel experiments. Bearing this caveat in mind, the MLP achieved successful classification of 40 of the 44 test images, a success rate of about 91%, which is higher than for any of the 128 × 128 experiments.

Further experiments, which involved pre-processing the images in a variety of ways (see [25] for details), training MLPs on the preprocessed images, and averaging the output from several MLPs when applied to the test set, resulted in a highest classification rate of about 98%, or 43 of the 44 test images. This approach is similar to the use of neural net ensembles [26], in which several nets are trained from different initial conditions. During training they explore different areas of the error surface, potentially converging at different points in weights space, and therefore having different generalization properties. The outputs of the MLPs are combined by averaging or voting. This should result in higher performance with fewer spurious errors.

14.5.2.3 Discussion of the single resolution approach

Although they show promise when compared to conventional techniques such as template matching, single-resolution MLP feature classifiers in all cases achieved detection rates of less than 100% on the relatively small test set used. In order to improve performances modestly, significantly increased resources (e.g. neural net ensembles) are required. In view of these diminishing returns it seems unrealistic to strive to produce a system capable of 100% recognition rates by this approach.

In the higher-resolution work, the nets were confined to a relatively small search window in the locality of the feature to be detected (the left eye) during both training and testing. If a trained net is scanned across the whole of a higher-resolution image spurious detections, which were referred to in the context of the lower-resolution work, become a problem.

In addition, the computation required to scan an MLP over images of the size used in the highest-resolution experiments equates to about 500 million multiply/add operations. Thus an eventual implementation capable of real-time feature location in sequences of many frames per second would currently require much expensive high-speed hardware.

14.5.3 The multi-resolution approach

In order to overcome some of the limitations of single-resolution MLP feature detection described in section 14.5.2, a multi-resolution approach can be adopted. In this approach, shown schematically in Fig. 14.13, a low-resolution version of the image to be searched is generated by filtering and subsampling. MLP feature locators are scanned across the low-resolution image to generate candidate search areas for each type of feature. The address of a pixel in the low-resolution image flagged as being a candidate feature location is passed to a search supervisor, where it translates into an area of pixels in the high-resolution image. This high-resolution search area can be scanned by an MLP feature detector to determine if the feature is in fact present, and to find its precise location.

Provided that the number of candidate search areas is small, this scheme requires much less computation than scanning the entire high-resolution image. However, as it stands it will still be prone to the location of spurious features, and will fail to locate features that are present. These problems

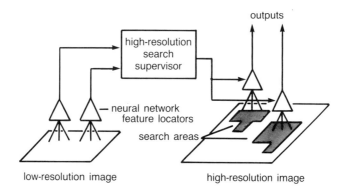

Fig. 14.13 Multi-resolution approach to feature location.

are ameliorated by higher-level processing using two techniques, both of which exploit knowledge about the source image. In the first, intra-frame knowledge about the relative positions of features is used. The image is assumed to be a single human head-and-shoulders image; therefore simple rules such as 'left eye is to the left of right eye' and 'mouth is below eyes' can be used to select likely triples of candidate feature points. The second technique to improve performance is to use inter-frame knowledge. Constraints can be put on the extent of allowable motion between frames, and if a feature is not detected in a particular frame, its position can be extrapolated or predicted from data obtained from previous frames. If the probability of locating a feature in each frame is 0.9, then in a sequence of three frames the probability of locating the feature in at least one of these frames is 0.999, assuming that the probabilities are independent (an assumption which may well be questioned!).

A prototype multi-resolution system has been developed for tracking features (left eye, right eye, mouth) in sequences of human head-and-shoulders images. It embodies higher-level processing exploiting knowledge about the source image and is known as the hierarchical perceptron feature locator (HPFL). It operates on source images of resolution 256×256 pixels, creating a 16×16 low resolution image by pixel averaging and sub-sampling. The component parts of the HPFL, shown schematically in Fig. 14.14, are now described.

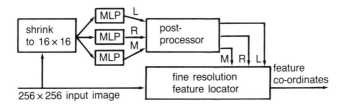

Fig. 14.14 Block diagram of a hierarchical perceptron feature locator incorporating post-processing of search areas generated at coarse resolution. Images containing search areas for left eyes, right eyes and mouths are labelled L, R and M respectively.

14.5.3.1 Obtaining candidate feature points in the low-resolution image

The feature detectors required in the low resolution stage of the HPFL are required to perform a different function from the feature detectors described in section 14.5.2, or those required in the high-resolution stage. The purpose of the low-resolution feature detectors is to locate all candidate feature points; therefore the detection of a point which is not a feature point (a false positive) is much less serious than the failure to detect a point which is a feature point (a false negative). False positives can be pruned out by various methods at a later stage, but false negatives present a more serious problem. This conditions the way the detectors are trained, as will be described.

The use of a single-layer perceptron to detect eye positions in 16×16 pixel images has shown some promise (see section 14.5.2.1). Its performance is perhaps surprising in view of the fact that its decision surface is a single hyperplane in 25-dimensional pattern space. In order to produce more complex decision surfaces, an MLP with a number of hidden nodes should be used. However, there is usually a trade-off between number of hidden nodes and ability to generalize. In order to achieve a balance between ability to generalize, and ability to form more complex decision surfaces, 'second-degree' neurons have been used.

The output of a second-degree neuron (a second-degree neuron with two inputs is shown schematically in Fig. 14.15(a)) is given by

$$y_j = f(w_0 + w_{ij}.x_i + v_{ij}.x_i^2) \qquad (1)$$

where v_{ij} are the weights associated with squared inputs. The more common linear neuron function can be obtained by setting these weights to zero. The second-degree neuron is a small step in the direction of higher order neural nets (HONNs) [27], and can form a richer set of decision surfaces than the linear neuron. A single second-order neuron can form a closed decision surface (a hyper-ellipsoid) independent of the dimensionality of the pattern space, whereas it requires a two-layer MLP with a minimum of $N+1$ hidden linear neurons to form a closed decision surface in N-dimensional pattern space. Assuming an infinite pattern space, a closed decision surface is one that encloses a finite region of pattern space. An open decision surface encloses an infinite region of pattern space. Figures 14.15(b)-(d) show three possible forms of decision surface ((b) simple open, (c) closed and (d) disjoint open) that a single second-order neuron can form.

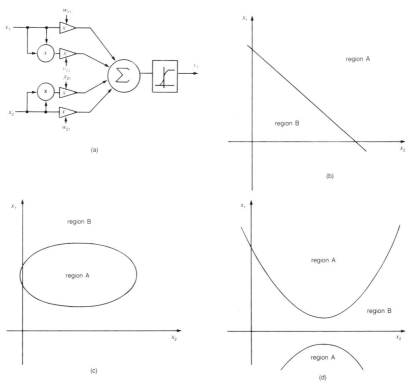

Fig. 14.15   (a)   Two input second degree   (b)   Simple decision boundary
                (c)   Closed elliptical   (d)   Disjoint open decision boundary.

Because the squared terms can be generated by pre-processing the input pattern vector, nets with second degree neurons in the first layer can be trained using the conventional backpropagation algorithm.

MLPs with a 5×5 input window, two hidden second-degree neurons and a single linear output neuron have been used as detectors to perform candidate feature point classification for the HPFL. Each detector is scanned across the 16 × 16 pixel image, and its thresholded output is used to create a 16 × 16 binary image known as the search area image. Unlike the earlier low-resolution experiments, separate detectors are used for left and right eyes. There are therefore three search area images, one each for the left eye, right eye and mouth. After further processing (to be described later) these are passed to the high-resolution stage, where they are used to define search areas for the feature detectors in the high-resolution image.

As the MLP is scanned across the source image there is one occasion when the scanning window is closest to the desired feature. When this occurs the expected output of the MLP is a high value (1.0), and the pattern vector on its input will be referred to as a feature vector. In all other positions the MLP output is expected to be low (0.0). The pattern vectors that form the input to the MLP on these occasions will be referred to as background vectors.

The MLPs were trained on 30 images selected from the set of 60 images used in the experiments described in section 14.5.2. For each low-resolution image there are 144 MLP input window positions, corresponding to 143 background vectors and one feature vector. If each vector is presented only once during a training epoch then the contribution to weight updating made by the feature vector is swamped by the effects of the background vectors; failure to detect the feature vector means that only 1 in 144 pattern vectors are misclassified, a success rate of 99.3%! This problem is avoided by presenting the two classes of pattern vector in a 1:1 ratio; each time a background vector is presented to the MLP it is followed by the feature vector from the same image.

Because of the need to ensure that all feature vectors are detected, whilst minimizing the number of background vectors that result in false positives, a selective training procedure was used. In this procedure a pattern vector is used only if the MLP's current response to it is considered unacceptable. In the selective training procedure the feature vectors are presented to the net at the end of a training epoch. If any feature is misclassified, backpropagation is applied for that vector. This is repeated until all feature vectors are correctly classified. If all feature vectors are correctly classified then in the next epoch all pattern vectors are presented to the net, but backpropagation is only applied for those that are misclassified. Misclassification is deemed to have occurred if the net gives an output <0.5 for an expected output of 1.0, or if the output is >0.5 for an expected value of 0.0.

The results of testing the three feature detectors on the 30 images of the training set and on 32 frames of a head-and-shoulders sequence of a subject not in the test set are shown in Table 14.1. The number of rejections (failure to detect a feature vector) must be read in conjunction with the figure for

330 ARTIFICIAL NEURAL NETS

mean search area, which is a measure of the number of false positives. For good net performance both of these figures should be low; it is perfectly possible to have no rejections with a search area of 100%, but this is obviously undesirable.

Table 14.1 Test results for HPFL low-resolution stage (REJ = number of rejected features, A,% = percentage of search area retained).

	Raw outputs pruning				After spatial pruning				After temporal expansion				After pixel			
	Train		Test		Train		Test		Train		Test		Train		Test	
	REJ	A,%	REJ	A,%	REJ	A,%	REJ	A,%	REJ	A,%	REJ	A,%	REJ	A,%	REJ	A,%
left eyes	0/30	4.45	0/32	2.19	0/30	1.09	2/32	0.62	—	—	2/32	0.60	—	—	0/32	4.08
right eyes	0/30	0.74	3/32	0.54	0/30	0.61	3/32	0.48	—	—	3/32	0.44	—	—	0/32	3.59
mouth	0/30	1.54	19/32	2.16	0/30	0.90	19/32	0.99	—	—	19.32	0.76	—	—	0/32	4.00

14.5.3.2 Exploiting knowledge to prune the search space

As stated earlier, for the low-resolution stage to perform well, it is necessary to eliminate false negatives whilst keeping the number of false positives low. The function of the low-resolution stage post-processor is to reduce the number of false positives, whilst retaining the true feature points. It has three consective stages: spatial pruning, temporal pruning and pixel expansion.

Spatial pruning
Spatial pruning is an intra-frame process that exploits geometric constraints imposed by facial structure on eye and mouth feature locations.

Consider a triple combination of candidate feature points, one point being taken from each of the search area images generated from a single frame. The pixels have co-ordinates r_L, r_R, r_M for candidate left eye, right eye and mouth positions respectively, as shown in Fig. 14.16. Some combinations are clearly not feasible, because they would imply, for instance, that the viewpoint is behind the head. This would happen if the candidate left eye position is to the right of the candidate right eye position. Other combinations might imply an unnaturally distended face. A set of geometric feasibility criteria can be established, and each possible triple that can be generated from the search area image can be tested to establish whether it satisfies these criteria. If not, that triple is rejected; this is the basis of the spatial pruning algorithm.

The parameters used in the HPFL low-resolution stage post-processor to establish a set of criteria are shown in Fig. 14.16. L is the spacing between the candidate eye points and θ is the angle that L makes with the horizontal. (U,V) are the co-ordinates of the candidate mouth point with respect to a frame of reference that has as its principal axes the line L, and an orthogonal line intersecting it at its midpoint. The following inequalities are tested:

$$L_{min} \leq L \leq L_{max}$$

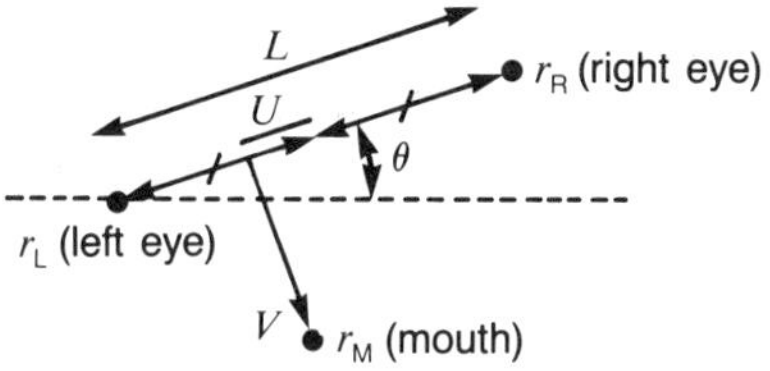

Fig. 14.16 A configuration of candidate eye and mouth points in a coarse image, showing the parameters involved in hypothesis testing by the spatial pruning algorithm (L,θ,U,V).

This implies that all images are at approximately the same scale, and restricts head rotation.

$$|\theta| \leq \theta_{max}$$

This restricts the range of orientations in the image plane.

$$(U/L) \leq (U/L)_{max}$$

This restricts the mouth position to lie approximately on the perpendicular axis through the midpoint between the eyes.

$$(V/L)_{min} \leq (V/L) \leq (V/L)_{max}$$

This rejects unnaturally compressed or distended face shapes.

The maximum and minimum allowed values are calculated from the training set of images, and relaxed to allow for greater variability in test images. Pixels in the search area image that do not belong to at least one triple that satisfies the above criteria are considered to be false positives, and are deleted.

The spatial pruning algorithm was applied to the images in the training and test sets. If all triples in a particular image failed to meet the pruning criteria then pruning was deactivated for that image. When pruning was not deactivated, the retained pixels in all cases included all the candidate feature points corresponding to correct location of the feature. The average reduction in search area over each data set was over 40%. The results are presented in Table 14.1.

Temporal pruning

The occurrence of images containing no valid triples indicates the presence of one or more false rejections. The spatial pruning algorithm can therefore be used as a method of detecting false negatives. When false negatives are detected in one frame of a sequence, data from previous frames can be used to estimate the position of rejected feature points, and to allow some 'spatio-temporal' pruning.

The above scheme has not yet been implemented in the HPFL. Currently the only form of temporal pruning implemented is an inter-frame process that imposes motion constraints on candidate feature points. By comparing locations of candidate feature points in adjacent frames, false positives can be rejected. Candidate feature points are constrained to move by no more than one pixel position per frame in any direction. For a 25 frame/s sequence this corresponds to about 1.5 image widths per second, which is not very restrictive.

Temporal pruning is applied after spatial pruning. Search area images of the same type are compared for adjacent frames of a sequence. All candidate feature pixels in the current frame that are more than one pixel away from a candidate feature pixel in the previous frame are deleted. This has the effect of reducing the search areas by removing spurious false positives which occur randomly.

The effects of temporal pruning are shown in Table 14.1 for the test set data (temporal pruning cannot, of course, be applied to the still images of the training set).

Pixel expansion

In the final post-processing stage, pixel expansion, the remaining candidate feature points are expanded from single pixels to an array of 3×3 pixels. Although this actually increases search areas, it is necessary because the generation of a coarsely subsampled image can give rise to sampling problems, where the desired feature in the high-resolution image lies across the boundary of two or more low-resolution pixels. In such cases a false negative may occur at the feature point, with a false positive in an adjacent pixel location. The false positive survives the spatial and temporal pruning processes, being close to the location of the feature point.

Pixel expansion ensures that the true feature point is brought within the search area, at the cost of increasing that search area. Table 14.1 shows the effect of pixel expansion on false negatives and mean search area, for the 32 frames of the test set.

14.5.3.3 Testing the low-resolution stage

As described above, the low-resolution stage of the HPFL has been tested using a 32 frame sequence of a female subject. The subject in the sequence did not appear in the 60 still images used to train the feature detectors. Figure 14.17(a) shows the initial search area images produced by the MLP feature detectors, superimposed on an example from the sequence. Figure 14.17(b) shows the results of spatial pruning; this has reduced the number of candidate triples, and hence the area that has eventually to be searched at the high resolution stage, from 56 to 12. Temporal pruning, shown in Fig. 14.17(c), reduces the number still further, to six. This shows the feature point corresponding to the mouth, which was adjacent to a candidate feature point prior to expansion, being absorbed into the search area.

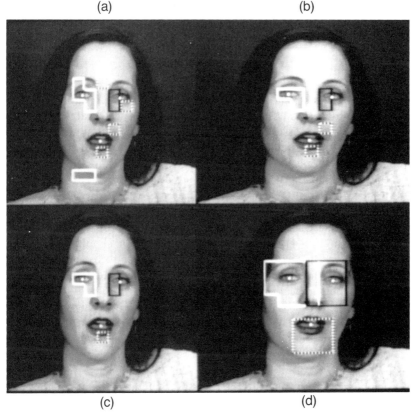

Fig. 14.17 Search area generation for eyes and mouth in an example frame at different stages of processing: (a) before post-processing, (b) after spatial pruning, (c) after temporal pruning and (d) after expansion. Boundaries of search areas are white for the left eye, black for the right eye and hatched for the mouth. Exact locations of features are shown as white dots.

14.5.3.4 Feature location in the high-resolution image

After pixel expansion, the search area images are passed to the supervisor of the high-resolution stage of the HPFL (see Fig. 14.14). Each pixel in a search area image corresponds to a 16×16 block in the high-resolution image. The high-resolution feature detectors are only scanned across areas of the high-resolution image corresponding to a candidate feature pixel in the search area image.

Currently, only a left eye detector has been implemented for the high-resolution stage. There is scope for post-processing of the detector outputs at the high resolution stage, to perform both intra-frame and inter-frame verification. This has not yet been implemented. Nevertheless, when the high-resolution detector was run on the output of the low-resolution stage for the test sequence described in section 14.3.3.3, all left eyes were located to within two high-resolution pixels.

14.6. CONCLUSIONS

Neural nets, and particularly MLPs, have several useful applications in visual telecommunications. A description of the principles of neural nets has been given, and areas of possible application in visual telecommunications have been discussed. In particular their use in object location and classification, where they have applications in intelligent surveillance, and in low bit-rate videotelephony, has been more extensively described.

As a case study in the application of neural nets to the latter problem a hierarchical perceptron feature locator has been described in some detail. The HPFL, as described in this chapter, is a method of locating features in human head-and-shoulders image sequences that should prove to be relatively robust. It needs extension to include error recovery in the post-processors, high-resolution right eye and mouth detectors, and perhaps intra- and inter-frame verification at high resolution.

The system currently has most difficulty with people who are wearing glasses. In retrospect we would perhaps be justified in excising these examples from our data set, in much the same way that data of poor subjective quality is sometimes removed from speech data sets. The problem of people with glasses could be dealt with by specific 'glasses' recognizer systems.

The underlying principles of the HPFL system mean that it could be used to perform feature location and tracking on a range of sequences of restricted scens, by customizing its various modules. There are a number of ways in which the system could be evolved further. At present it deals with images at an approximately fixed scale, and assumes that there is only one head and shoulders in the image. Possible extensions include allowing the system to operate over a range of scales (possibly by using a multiresolution pyramid), and allowing images that have a number of subjects in them. Spatial and temporal pruning, which are currently knowledge-based, could be implemented by neural techniques. This is currently under investigation.

ACKNOWLEDGEMENTS

A number of people have contributed to the work described here. We would like to acknowledge the contribution of Bill Welsh, who performed some of the early work which lead to the development of the HPFL system, and Graham Seabrook for the short section on applications of neural nets to intelligent surveillance.

REFERENCES

1. Welsh W J: 'Model-based image coding', Br Telecom Technol J, 8, No 3 (July 1990) and Chapter 11 of this book.

2. Rosenblatt F: 'Principles of Neurodynamics', Spartan, New York (1962).

3. Minsky M L and Papert S A: 'Perceptrons' MIT Press, Cambridge, Mass (1969).

4. Rumelhart D E, Hinton G E and Williams R J: 'Learning internal representations by error propagation', in 'Parallel distributed processing', 1 Eds Rumelhard D E & McClelland J L, MIT Press, Cambridge, Mass (1986).

5. Kohonen T: 'Self organisation and associative memory', Springer-Verlag, Berlin (1984).

6. Tattersall G D, Linford P W and Linggard R: 'Neural arrays for speech recognition', Br Telecom Technol J, 6, No 2, pp 141—163 (April 1988).

7. Aleksander I, Thomas W V and Bowden P A: 'Wisard: a radical step forward in image recognition', Sensor Review, 120—124 (July 1984).

8. Fukushima K, Miyake S and Ito T: 'Neocognitron: A neural network model for a mechanism of visual pattern recognition', IEE Trans Systems, Man and Cybernetics, SMC-13, No 5 (Sept/Oct 1983).

9. Cotrell G W, Munro P and Zipser D: 'Learning internal representations for grayscale maps: an example of extensional programming'.

10. Kaouri: 'A comparison of the LBG algorithm and neural network based techniques for image vector quantisation', Report on work carried out as a short-term research fellow at BT Laboratories (July-September 1989).

11. Linde Y, Buzo A and Gray R: 'An algorithm for vector quantiser design', IEE Trans Comm COM28, No 1 (Jan 1980).

12. Wright M J: 'Training and testing of neural net window operators on spatiotemporal image sequences', in 'Neural Networks in Vision, Speech and Natural Language', Eds Nightingale C and Linggard R, Chapman and Hall, London (1991).

13. Fischler M A and Firschein O: 'Readings in computer vision: Issues, problems, principles and paradigms', Morgan Kaufmann, Los Altos (1987).

14. Jain R N and Binford, T O: 'Ignorance, myopia and naivete in computer vision systems', CVGIP: Image Understanding, 53, No 1, pp 112—117 (Jan 1991).

15. Marr D: 'Vision', Freeman, New York (1982).

16. Houghton A, Hobson G S, Seed L and Tozer R C: 'Automatic monitoring of vehicles at road junctions', Traffic Eng and Control, pp 541—543 (Oct 1987).

17. Dickinson K W and Waterfall R C: 'Image processing applied to traffic', Traffic Eng and Control, 25, No 1 pp 6—13 (1984).

18. Seabrook G R: 'Real-world object recognition with a neural network', Proc 6th Intl Conf on Image Analysis and Processing (Sept 1991).

19. Persoon E and Fu K: 'Shape discrimination using Fourier descriptors', IEEE Trans Systems Man and Cybernetics, SMC-7, No 3, 170—179 (March 1977).

20. Hutchinson R A and Welsh W J: 'Comparison of neural networks and conventional techniques for feature location in facial images', Proc 1st IEE Int Conf on Artificial Neural Networks, London (October 1989).

21. Bishop J M and Torr P: 'The stochastic search network' in 'Neural Networks in Telecommunications', Eds Nightingale C and Linggard R, Chapman and Hall, London (1991).

22. Debenham R M and Garth S C J: 'The detection of eyes in faces using radial basis functions', in 'Neural Networks in Telecommunications', Eds Nightingale C and Linggard R, Chapman and Hall, London (1991).

23. Allinson N M and Johnson M J: 'Digital realisation of self-organising maps for image recognition', in 'Neural Networks in Telecommunications', Eds Nightingale C and Linggard R, Chapman and Hall, London (1991).

24. Hines E and Hutchinson R A: 'Application of multilayer perceptrons to facial feature location', Proc 3rd IEE Int Conf on Image Processing and its Applications, Warwick UK (July 1989).

25. Nightingale C and Hutchinson R A: 'Artificial neural nets and their application to image processing', Br Telecom Technol J, 8, No 3, pp 81—93 (July 1990).

26. Hansen L K and Salamon P: 'Neural network ensembles', IEEE Trans on Pattern Analysis and Machine Intelligence, 12, No 10 (October 1990).

27. Maxwell T et al: 'Transformation invariance using high order correlations in neural net architectures', IEEE Int Conf on Systems Man and Cybernetics, Atlanta, Georgia (October 1986).

Index

AALs (ATM adaption layers) 173–4
ACT (audio conference terminal) 106, 108
Affine transforms 268–9, 271–5, 276–7
ANNs (artificial neural nets) 300, 304, 306
ASICs (application-specific integrated circuits) 94, 146, 148
ATM (asynchronous transmission mode) 21, 22, 23, 129, 156, 158, 159, 161, 173–5
Audio, multiplexing with other services 121
Audio information signals 118–19
Audiographic teleconferencing 17
Audiovisual services 19, 21, 23, 25, 100–11, 113–18
Axons 302

BAS (bit allocation signals) 108, 124–6
Base layer, codec data 164
Baseband protection ratio 65
BC (bearer capability) service 129–30
BCH (Bose Chaudhuri Hoquenghem) technique 161
BER (bit error ratio) 63
Best-match block search 223
Big star local distribution 39
Bit interleaving 72, 73
Block matching 280
Boundary coding 272
Boundary conditions, snakes 256
Broadband networks, videophony 95–6
BSS (broadcast satellite systems) 45

Call control 129–31
Cameras, videophony 88–91
CANDIDE wire frame model 198, 205–6
CCIR (International Radio Consultative Committee) 27, 29, 31

framework for audiovisual recommendations 116–18
H.261 recommendations 133–6, 137–41, 144–9, 153, 157, 159, 182, 189
multipoint recommendations 104–5, 108, 109
CCITT SGXV Specialists Group for visual telephony 179–80
CD-ROM 178, 179
Charges, videophony 91
Cholesky decomposition 259
CIF (common intermediate format) 135–7, 142, 144, 145, 146, 185
Clock recovery, codecs 174
Codebooks, for MBC 198–205
Codecs 6, 8, 21, 102, 120, 127
    fixed-rate networks 156–7
    low bit rate 133–54
    VBR 161–74
    videophones 92, 93, 94
Coding
    binary algorithm 150–3
    efficiency 168
    fractal techniques 271–9
    Huffman 194
    moving-picture 290–5
    still-pictures 279–91
    videophone 150–4
    see also Codecs
Collage theorem 272
Companding laws 92–3
Compression, algorithms for 195
Conversational services 13, 16, 24
CYCLOPS system 102

D-MAC, BER 63
Database services 13, 16, 18, 22, 24
DATV (digitally assisted TV) 48, 49, 63
DBS (direct broadcasting by satellite) 45, 55–6, 60–1
DCT (discrete cosine transform) 8, 10,

57, 65, 134–5, 137–41, 145, 147, 181, 182, 183, 186
Delaunay triangulation 208
Delta rule, neural nets 305, 306
Digital multiplexing, videophony 97
Digital PBXs 20
Digital TV 29–30
Digital video coding for storage 180–92
Discrete cosine transform, see DCT
Displays, videophony 86–7
Distribution services 13, 16
Domain blocks 277–8
Double star distribution 39–40
DPCM (differential pulse code modulation) 33, 57, 134–5, 194, 291–2
DSPs (digital signal processors) 76–7, 93–4, 142, 143, 147–8, 152
Dumbell audiovisual networks 103–4
DVI (digital video interactive) 178

Edge-detection, see Snakes
EDTV (extended-definition TV) 53
Enhancement layer, codec data 164–7
Error protection and correction 72–3, 73–5, 125, 145, 159, 161
ESPRIT EC programme 106
ETS (European Telecommunications Standards) 108
Eye
  perception of 6, 296
  resolution of 2
Eye detection 212–19, 320, 321–4
Eye-contact, videophony 90

Facial image analysis 199–201
FACS system, facial movements 198
Facsimile information signals 120
FAS (frame alignment signals) 108, 123, 125, 126
FDM (frequency division multiplex) 41, 55
Feature codebook generation 203–4
Feature detection, neural nets 319, 321–5, 327–35
Feature location, faces 210–22
Feature tracking, for MBC 202–3
Flicker, picture 28, 89, 136
Format, pictures 135–7
Formats, MAC and HDMAC 46–8
Fourier descriptors 317

Fractal data compression 266–97
Frame dropping, redundancy reduction 11
Frame structure for multiplexing 121–3
FSS (fixed satellite systems) 45

Galerkin's method 253
GOB (group of blocks), codecs 137, 141
Grey-scale algorithm 153–4

Hardware, codecs 173–4
HDMAC 37, 45–69, 59–61, 62–5
HDP (high-definition progressive) technique 32, 65
HDTV (high-definition television) 3, 26, 30–4, 36–8, 45–69, 192
HLC (higher-layer capability) 129–30
HONNs (higher order neural nets) 327–8
HPFL (hierarchial perceptron feature locator) 319, 326, 327–35
HSM (horizontal symmetry measure) 213

Image compression 6–11, 37, 318
Image identification and compression, neural nets 310–18
Image processing 19
Imps (subjective impairment units) 60, 68
IMTRAN system 102
Information transfer services (between machines) 13, 16
Interference, see Noise
Interframe coding 10
Interleaving 28
IPAs (interface protocol adapters) 107–8
ISDN (integrated services digital networks) 17, 20–2, 38, 85–6, 91–4, 97, 118, 121, 129–30, 133, 156

Jack in the Box video test sequence 163–4, 167–72
Jacquin's decoder 281, 284–9
Jitter, codec cells 174

K (Kohonen) nets 308–12, 320, 322
K-rating 61–4, 68–9

INDEX 339

KBC (knowledge based coding) 194

LANs (local area networks) 22, 86, 96
Lawson's swapping algorithm 208–9
Learning function, neural nets 310
Least squares estimation 224–5
Lighting, videophony 88–9
Line shuffling 51
LLC (lower-layer capability) 129–30

MAC 45–6, 46–8, 56, 59–61, 65
  K-rating 62
Mapping, wire frames 235–6
MBC (model based coding) 194–244, 245
MCU (multipoint control unit) 102, 104–5, 106, 127–9, 130
Mesh audiovisual networks 102–3
Messaging services 22
MIAC (multipoint interactive audiovisual communication) project 106–9
MIAs (multipoint interactive audiovisual system) 109–11
MLP (multilayer protocol), telematic signals 121, 127
MLPs (multilayer perceptrons) 211, 212, 214–15, 304–7, 311–12, 320–5, 328–9
Model based coding, see MBC
Model conformation 229–42
Models, 3D 196–8
Motion compensation 10–11, 139–40, 145, 188–91
Motion estimation 222–9
MPEG (Moving Picture coding Experts Group) 179–86, 189–93
MPTV (moving-picture television) 14, 17, 119
Multiplexing information flows 121–3
Multipoint conference types 104
Multipoint connections 127–9, 130
  see also MCU
MUSE 37, 45–6

Nagao's algorithm 202, 212, 230
Neocognitron 310
Network loading, simulation 170–2
Network policing 159–60
Networks, multipoint audiovisual 102–4
Neural computing 300–4

Neural nets, artificial 299–335
Neuron
  biological 303
  second degree 327
NICAM (near-instantaneously companded audio multiplex) encoders 71–84
NMSE (normalized mean square error) fidelity measure 195
Noise 54, 55–6, 56–7, 58–9, 67
  vision scrambling 64
  wideband random 59–61
Nose and mouth location 217–19

Object identification 315
Optical fibre transmission 36, 38–42
ORATOR system 101
Orwell protocol 162
OSI (open system interconnection) 114

PAL system 28, 29, 46–8, 53–4, 55–6
Passive bus architecture 40
Pattern recognition 314
PCM (pulse code modulation) 33, 57, 95, 119, 184, 185, 186
PDM (pulse density modulation) 33
Peano curve 266–7
Perceptron, see HPFL
Phong shading 197
Piecewise linear mapping 208
Pixel expansion, feature detection 332
PON (passive optical network) 41–2
Prediction techniques 7, 10, 140, 196, 291
Processing services 13, 16, 18
Progressive scanning 31, 185
PSTN (public switched telephone network), videophony 85–6, 96–8, 150–4
Psycho-visual effects 296
PTDs (parallel transfer disks) 134

QCIF (quarter common intermediate format) 136–7, 144, 145, 147, 148–9
QPSK (quadrature phase shift keying) 75
Quad-tree compression process 151–2, 194
Quality/cost, HDMAC/PAL 52–6

RACE optical fibre projects 38, 42
RAM 134, 148
Recognition, object classes 12
Region coding 271–2
Resolution
  TV image 3–5
  videophony 95
Rotational procedures 223

S/N (signal/unweighted noise) ratio 66–7
SBM (symmetry blob measure) 214
Services, audiovisual 12–15, 113–18
Shannon's information limit 296
Shape testing 211–12
Simulation, video codecs 162–72
Snake segment location method 220–2
Snakes 210–11, 215–17, 245–64
  closed loop and fixed end 230–5, 254–8
  finite elements and differences 250–8
  properties 246–9
  system solution 258–9
SNR (signal-to-noise ratio) 167–70
Soft facsimile 14
Spatial pruning, feature detection 330–1
Speech, and lip movement 11
SPTV (still-picture television) 12, 17, 101–2, 119
Star audiovisual networks 103
Static image analysis 210–22
Still picture coding 279–90
Stochastic nets 320
Surveillance services 13, 16, 18, 22
Synapses 302
Synthetic sequence tests 226–9

TA (terminal adapter) 173
Tariffs
  network loading effects 159
  systems and networks 23–4
  VBR video 175
TDM (time division multiplex) 41
Teleconferencing, *see* Audiovisual services
Teleseminas 17, 22
Template matching techniques 322–3
Temporal pruning, feature detection 332

Terminals 113–15, 124–6
Text information services 121
Texture mapping 209–10
3D coding 291–2
3D modelling 196–8, 205–10
Traffic monitoring, neural nets 316–17
Transaction services 13
Transform coding 7–10, 33, 194
Transmission errors 19–20
TV
  chain quality evaluation 57–9
  digital, *see* Digital TV
  image quality 7
TV studios, digital equipment in 30

UMTS (universal mobile telecommunications system) 42

Variational problem, snakes 251
VBR (variable bit-rate) video 156–75
VCRs (video tape recorders) 177–8, 186
Video, multiplexing with other services 121–3
Video coding for storage, *see* Digital video coding for storage
Video data buffering 142–4
Video information signals 119–20
Video signal compression 104
Videoconferences, simulation sequences 162–4
Videophony 17, 85–99
  coding 150–4, 314
  hardware 93–4
  use of neural nets 318–19
Videotelephones, simulation sequences 162–4
Viewing angles 53
Vision, biological 315
VLSI (very large scale integration) 8, 94, 148, 181, 183
VSM (vertical symmetry measure) 213

Waveform distortion
  effect on quality 54
  linear, of luminance 61–4
WDM (wavelength division multiplex) 41
WISARD 310

ZELDA international test picture 283–5